NATIVE VEGETATION OF NEBRASKA

Native Vegetation of Nebraska

by

J. E. Weaver

UNIVERSITY OF NEBRASKA PRESS · LINCOLN

Library of Congress Catalog Card Number 65-13259

Manufactured in the United States of America

Contents

1. Location, Topographic Regions, River Systems, and a Little History 1

2. Deciduous Forest 13

3. Flood-Plain Forest 33

4. Lakes, Marshes, and Other Wet Land 41

5. Grasses of Bluestem Lowlands 55

6. Grasses of Uplands 68

7. Forbs of Lowlands 77

8. Forbs of Uplands and Viewpoints on Prairie 84

9. Soils 96

10. Transition to the Great Plains and Loess Hills 101

11. The Great Plains 113

12. Sand Hills 135

13. Evergreen Forests and the Northwest 155

14. Cultivated Crops of Grassland Soils 164

I.

Location, Topographic Regions, River Systems, and a Little History

Nebraska is in the central part of the prairie plains. Yet grassland is only one kind of its varied vegetation. A westward extension of the great deciduous forest occurs along the Missouri River. It is bordered by shrubs, patches of which occur in favorable places to the western border. The foothills of the northwest have a forest of pine not unlike that of the Black Hills. On the level lands and among the wind-blown hills of sand there are hundreds of lakes, swamps, and marshes. The Missouri, Platte, and Niobrara Rivers extend across the state. They and their numerous tributaries have built extensive flood plains with varied vegetation and often a border of woodland.

All of this vegetation has been extensively studied throughout half a century by the writer, his colleagues, and his students at the University of Nebraska. It was also explored by earlier botanists. The nature and beauty of this wonderful vegetation will be recorded simply and clearly so that both present and future generations may know about the native vegetation of their state (Figs. 1–5). A final chapter includes the development of cultivated crops in prairie soils, the depth of the soil at which they absorb water and nutrients, and their demands for water compared with those of prairie vegetation.

Why the vegetation is so diversified, where each kind occurs, and its relation to that of similar types in surrounding areas may well be considered before the vegetation itself is described. Nebraska is centrally located in the United States. It lies between the meridians of longitude 95° 25′ and 104° west of Greenwich and between parallels of latitude 40° and 43° north of the equator. Its extreme length is about 462 miles and it is approximately 200 miles wide. Its area, about 77,510 square miles, is greater than that of the New England States combined. Somewhat more than the eastern one-third lies in the Central Lowlands, which is an area of slight relief in the interior of North America and which extends far eastward beyond Iowa and Illinois. The remainder of the state lies in the

FIG. 1.—Typical forest of bur oak near Nebraska City. The trees are about 38 feet high. Gooseberry and buck brush are the principal shrubs.

central portion of the Great Plains Region. Altitude is 5,340 feet in Banner and Kimball Counties near the Wyoming line, where a few places are more than a mile above sea level. The Pine Ridge Area in the northwest has an altitude of about 4,000 feet. The state as a whole slopes eastward between 8.5 and 9 feet per mile. Consequently the rivers flow eastward or southeastward. Elevation in the northeast is about 2,000 feet but in the extreme southeast it is only 825 feet.

Mean annual precipitation, which is a very important factor in considering kind of natural vegetation, varies from 33 inches in the southeast to 23 inches in the northeast. Then it decreases westward to 17 and even 15 inches (Fig. 6). Other factors such as temperature, length of growing season, wind, etc., which affect climate will be considered with the vegetation types. The zonal soil groups from east to west—Brunizems, Chernozems, and Chestnut soils—which have developed under the prairie and plains climate and vegetation are best studied with the type of vegetation they support.

Fig. 2.—Flood-plain forest near Nehawka at the base of a steep north-facing slope. The trees are hackberry, white elm, red elm, green ash, and Kentucky coffee-tree. One very old hackberry on the riverside, not shown here, was more than 4 feet in diameter and 95 feet high.

The prairie of the eastern third of Nebraska is entirely representative of North American Prairie. This great prairie originally occupied all of the region lying east of the Great Plains vegetation of mixed prairie and the western margin of the great Deciduous Forest. The westerly border of the forest from Minnesota to Texas was mostly dominated by oaks and hickories and graded into grassland through a border of shrubs. This prairie, now generally called true prairie, is dominated mostly by big and little bluestem and might better be called the tall grass-mid (middle sized) grass prairie. It is bounded on the west throughout by the mixed prairie of the Great Plains, a mixture of mid and short grasses. The magnificent grassland of bluestems extended from southern Manitoba diagonally

Fig. 3.—Sand hills showing several small blowouts at Seneca in 1919. (From *Grasslands of the Great Plains.*)

Fig. 4.—View in one of the largest ungrazed prairies in eastern Nebraska. Lincoln, 1959.

FIG. 5.—A steamer breaking the native buffalo grass sod on the Box Butte Table in northwest Nebraska about 1915.

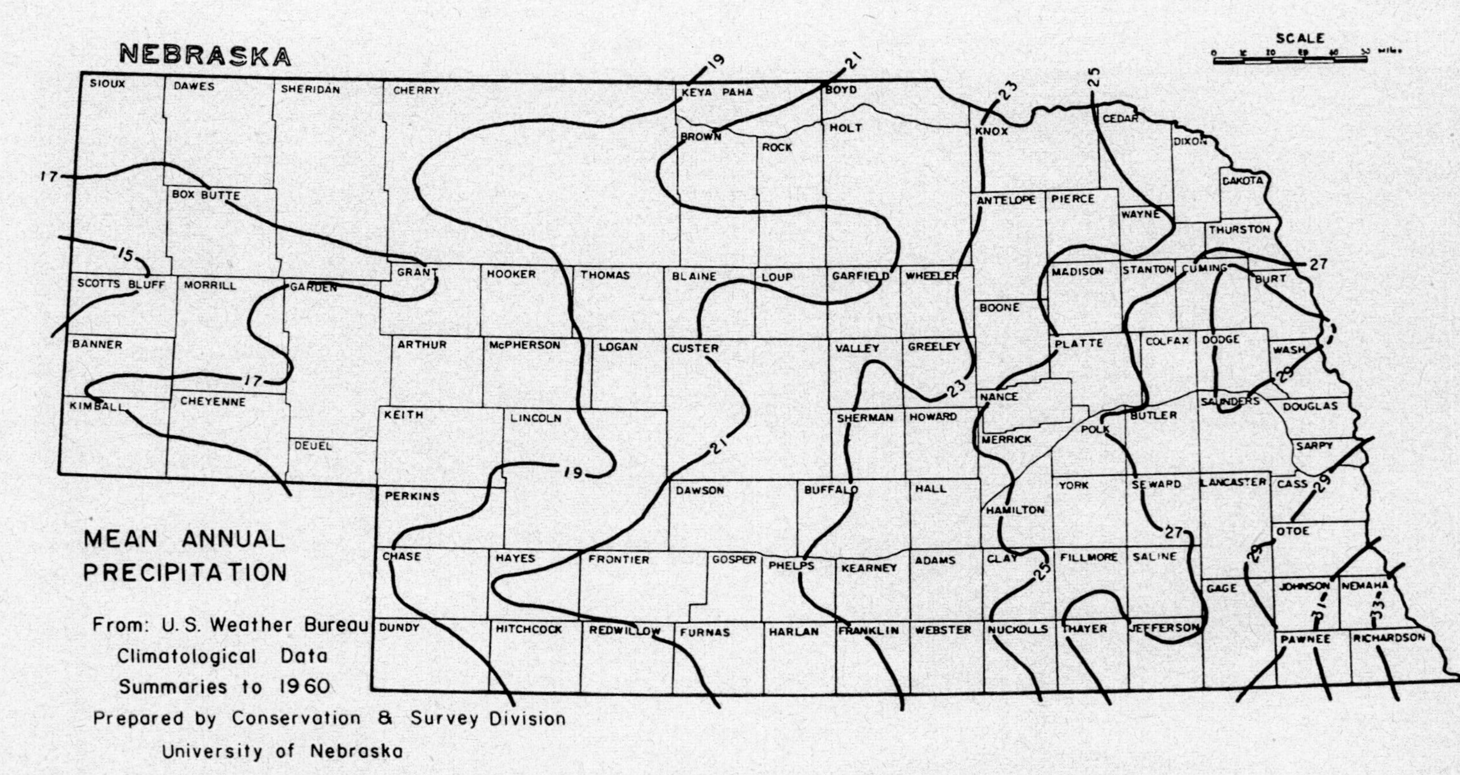

FIG. 6.–The counties of Nebraska and their precipitation in inches.

through Minnesota to southwestern Wisconsin and included the northern two-thirds of Illinois and part of western Indiana, with small areas in Ohio. It included northern and western Missouri and the eastern half of Oklahoma, for the most part, and extended far southward nearly through Texas. Iowa is the only state lying entirely within the domain of prairie. Nebraska is in the central part of it from north to south.

One study of more than a hundred representative areas of natural grassland early in this century revealed that the prairies of Nebraska were entirely representative of the general area. There was 100 percent agreement of the 11 major grasses in Nebraska and Ohio, and 87 percent of lowland forbs or non-grass-like herbaceous plants. Thus, throughout this wide area the carpet of prairie extended over level land, knolls, steep bluffs, rolling to hilly land, valleys, and extensive flood plains.

Some time may well be spent examining the topographic regions of the state (Fig. 7). While the state boundaries are purely political, except for the Missouri River, the Niobrara River on the north, the Republican River on the south, and the foothills on the west coincide approximately with these boundaries.

The Missouri River, with thousands of tributaries, drains the water from plains and prairie. It is ¼ to ½ mile wide where it begins to border northeastern Nebraska (Fig. 8). It continues to separate Nebraska from Iowa and Missouri for a distance of about 360 miles, but because of its sinuous course the actual length is approximately 500 miles. It is not only wide but in places it is 30 feet deep. It is quite swift, but has a fall of only about a foot per mile where it borders Nebraska. The river valley varies from a few to 17 miles in width. The valley is bordered by bluffs throughout most of its Nebraska course. In southeastern Nebraska the bluffs are clothed with a rather dense deciduous forest. The entire state is drained by the tributaries of the Missouri although some rivers, such as the Republican and Blue Rivers, first flow into Kansas. This great river is notable for the extremely shifting character of its tortuous channel. It rarely flows midway between the bluffs but moves from one side of the valley to the other, leaving first a broad stretch of flood plain on the Nebraska side and then cutting in beneath the overhanging bluffs. It has shifted its channel as much as a mile within a year. Near Omaha, for example, the river formerly flowed through cut-off or Carter Lake. But by straightening its course it left part of its channel in which cut-off lake is situated and thus transferred a part of Iowa to the west side of the river (Fig. 9).

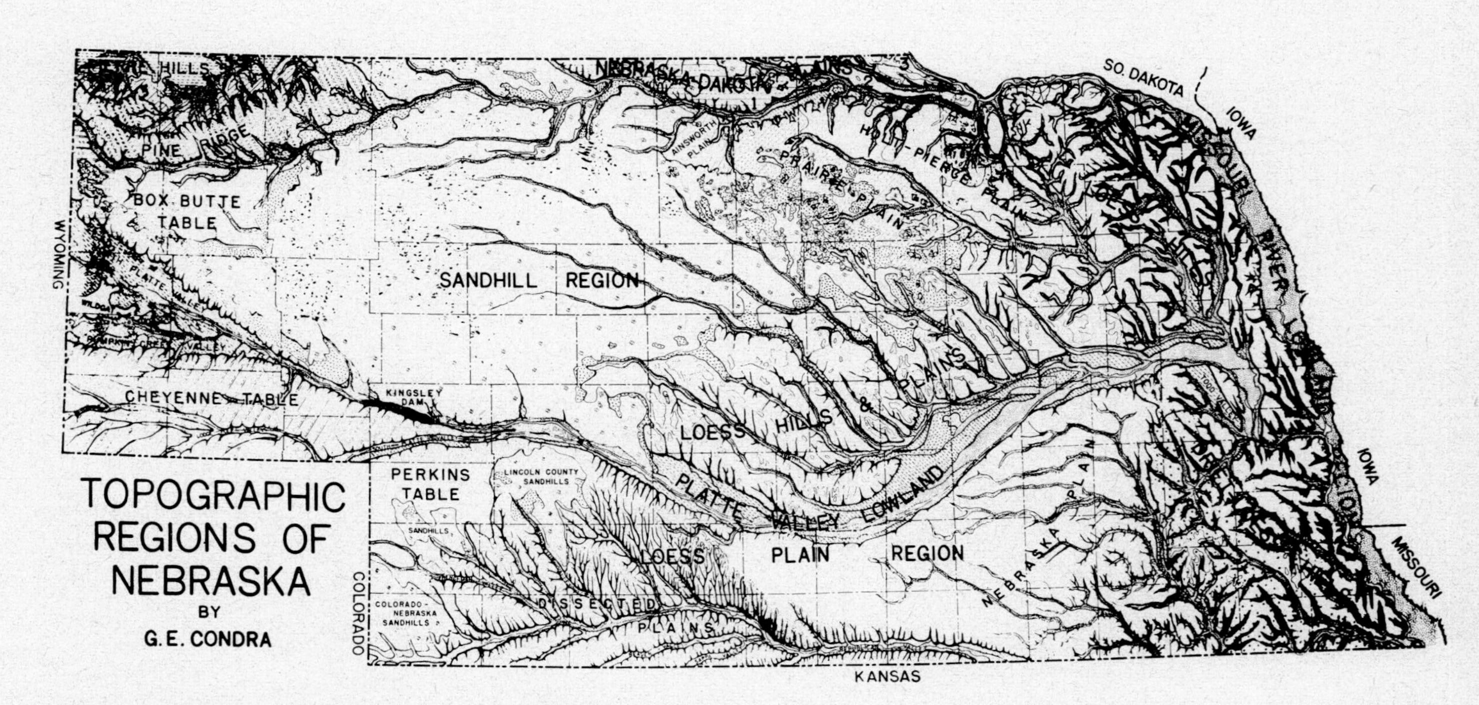

Fig. 7.—This is Nebraska; vegetation in the several regions will be described.

FIG. 8.—Missouri River opposite Vermilion, South Dakota. There is a young growth of cottonwood on the sandbar; a fringe of sandbar willow has been eroded away.

FIG. 9.—General view of a bluff facing northward along the Missouri River in northeast Nebraska, and a cornfield being washed away.

Most of the course of the Niobrara River, which rises in the high table lands of eastern Wyoming, is in northern Nebraska. It is a shallow but swift river with well-protected bluffs. It has a rather deep valley bordered by somewhat rough, stony land where it flows eastward down the slope nearly 400 miles without deviating more than 15 miles from the parallel of 42° and 40′. The flood plain is seldom more than ½ mile wide in Dawes County, where the main channel has a width of 25 to 40 feet. The bluffs of the Niobrara are steep and are covered with a dense growth of shrubs and small trees, and in some places the valley is heavily wooded. In the eastern portion of the state its banks are low and wooded. It is through this valley that many trees, shrubs, and some other plants common in eastern Nebraska extend far westward. Also along this route many species from the Black Hill Region have migrated far eastward.

The Platte River is the largest tributary of the Missouri in Nebraska. This river is formed by the union of the North Platte, which rises in North Platte Park of Colorado, and the South Platte, with its source in the South Park of Colorado. The Platte is broad, shallow, and usually heavily loaded with fine sand. In places it flows in a single channel but in others it consists of interlacing streams among sandy islands. In the last 20 miles of its course it becomes narrower and the valley slopes become steeper. Here wooded bluffs are quite pronounced. Rivers and streams are the highways along which woodland extends westward, often far into the state.

The Republican River rises in the sandy plains of eastern Colorado. In southern Nebraska it has a network of tributaries in the Dissected Plains. Like the Platte, it is shallow, usually heavily loaded with sand, and possesses a wide flood plain. The bluffs rise at some distance from the river and are generally bare. At their base lie rather broad, treeless valleys. Along the banks there is commonly a fringe of willows.

The Loup River is formed by the junction of the North, Middle, and South Loups. In the upper part of their course they lie in deep gorges with abrupt banks. They are among the finest streams in the West and are remarkable for their steady flow of pure, cold water. In their lower course they flow in narrow valleys covered for the most part with meadow land.

In general trees do not grow on the wind-swept prairies but along the larger streams or in well-watered ravines (Figs. 10, 11). Extensive deciduous forest occurs only in southeastern Nebraska, as a result of higher rainfall and the ameliorating effect of climate caused by a great river.

The various types of grassland, forest, and other vegetation in Nebraska can best be understood when something of their origin is known. This is explained in *Grasslands of the Great Plains,** from which certain paragraphs are given here.

FIG. 10.—Mowed prairie and bur oak forest near Seward. Here the trees grow only in the depressions between the hills.

FIG. 11.—Prairie and trees in a ravine on a north-facing hillside near Lincoln. Willow, boxelder, elm, sumac, and dogwood are the woody plants.

**Grasslands of the Great Plains, their Nature and Use* (Lincoln: Johnsen Publishing Company, 1956).

The work of students of the earth's structure and of plant and animal fossils of past geological periods, together with the study of past climates and climax vegetations, has given us the history of the earth's surface, its animal life, and its vegetation. The origin of mixed prairie probably dates back 25 million years to Tertiary times. In the Eocene period a warm temperate forest occupied the Great Plains. At that time the climate was warm and moist. But as the Rocky Mountains rose, they intercepted the moisture laden winds from the Pacific Ocean. Since water was precipitated mostly to the west of the mountains, only dry winds, which produced very little rainfall, reached the eastern side. Low summer precipitation was accompanied by dry winters.

During the Miocene period (which followed later), to the east of the Rocky Mountains increasing aridity reduced the forest vegetation. A wide extent of grasslands is indicated by the abundance and diversity of grazing animals. Indeed, grazing animals have roamed the Great Plains throughout geological ages. There were horses, camels, elephants, rhinoceros and primitive types of buffalo. They fed upon the grass and other vegetation. Their skeletal remains prove their existence. In Pleistocene time, about a million years ago, the ice of the glacial period covered only a relatively small part of the Great Plains. But the period of glaciation had nevertheless a profound effect upon the cover of vegetation. It seems certain that the Boreal Forest was pushed far southward and later retreated northward during warm, dry periods. Evidence of this is found in the persistence of white spruce to this day at higher altitudes in the Black Hills and the survival of paper birch and aspen, both cold climate species, in deep canyons and on elevated plateaus. Indeed many types of vegetation retreated and advanced across the Great Plains during the pulsation of cool, moist and warm, dry periods in Pleistocene time.

The eastern edge of the Black Hills has numerous relics of eastern Deciduous Forest trees, such as bur oak, ash, elm, and others, which here found refuge and remained behind the eastwardly retreating woodland which once covered the plains. A relic, in an ecological sense, is a community or a fragment of one that has survived some important change, often to become a part of the existing vegetation. Other evidence is the presence of hard maple and other remnants of Ohio forests in the deep Caddo Canyon in west-central Oklahoma. It is also believed that tall grasses occupied the Great Plains for a period following the retreat of the deciduous forest.

II.

Deciduous Forest

Despite the general grassland environment in Nebraska, the Missouri River has so greatly modified the climate along its course that a part of the deciduous forest extends some distance into the grassland. The deciduous forests of southern Ohio and eastern Kentucky grow under a mean annual rainfall of about 42 inches. In Nebraska the greatest average annual rainfall is about 33 inches. This is in the extreme southeast. Deciduous forest occurs westward across Missouri and also northwestward along the Missouri River in Nebraska. The number of woody species of considerable importance is more than 200 near the center of this great eastern forest but only about 80 in southeastern Nebraska. Of this number there is a further decrease of upland woody species to about 31 where the Missouri River first touches Nebraska. The nature and extent of this forest and its transition into grassland through a border of shrubs will now be examined.

The Missouri River has an average width of a little less than one-half mile where it reaches Nebraska and a valley varying in width from one-half to one and one-half miles. It is somewhat wider and deeper near the Kansas line and its valley is often very much wider. Near the Dakota line the valley is U-shaped and on an average about 150 feet below the summit of the first row of bluffs. But near the Kansas line it is broad and often 200 feet below the bluffs which border it on the west. This mighty river has not only eroded a bluff-rimmed depression, but numerous tributaries and their branches have also cut canyons below the general wind-swept prairie level. Thus the physiography of eastern Nebraska, and especially the southeastern portion, is such as to afford protection for tree growth.

A moderately high rainfall wets the soil, the large river-water surface adds humidity, and bluffs and canyons decrease wind movement; hence forest develops in an otherwise prairie region. In fact, forests and woodland, except in the extreme northwest, chiefly occur along the Missouri River and its tributaries. In the southeast, forest extends several miles from the river into the Drift Loess Hills. Extended studies reveal that, although forests have been destroyed by cutting except on the rougher land, the area of potential woodland had a width of approximately 25 miles in the extreme southeast, 5 miles in the vicinity of Omaha, but often less than ½ mile northward and westward.

Extensive examination of the deciduous forest reveals that it is composed of three different communities aside from the valley woodland along the small streams. They are named after the most important trees—the dominants—in each community.

Red Oak–Linden Community

This community is the most mesophytic. It occupies the best protected slopes and ravines along the rivers in the eastern part of the state. In the southeast it clothes all of the lower slopes and the north exposures of the higher ones (Figs. 12, 13). Linden and red oak are grouped because of their close topographical association, despite the fact that eastward linden is often a dominant in the maple-beech community and red oak in the oak-hickory community. In our forest they occur where water content of soil and humidity are highest and water loss through evaporation lowest. Many other trees might thrive here but they cannot grow in the dense shade produced by the two dominants. Linden usually occurs on the lower

Fig. 12.—Well-developed forest of red oak and linden on hilltops in southeastern Nebraska, six miles west of the Missouri River. Much land has been cleared of forest and is under cultivation.

FIG. 13.—Interior view of a forest of red oak east of Lincoln where the Weeping Water joins the Missouri. Nearly all of the pictures of forests were taken in 1928.

slopes, which are of higher water content, and red oak is found in the better drained areas. Examination of seedlings shows that red oak is better adapted to drier soil than linden. Roots of red oak reached a depth of 7 inches before the leaves began to unfold. Those of linden were 2 inches or less in depth. At the end of the second summer, red oak had a root depth of 6.5 feet and branch roots which spread outward about a foot on all sides of the plant. Those of linden, of a similar age, were only 2 feet deep. They were very well branched in the surface soil, but deeper branches were less than half a foot long. Ironwood, a small tree, is the only species that makes a good growth under the dense forest cover. Red oak and linden, which are larger trees than those of any other community, produce a thick stand.

This community is best developed in the southeast but extends up the Missouri River to somewhat beyond Ponca. Adjoining the flood plain it clothes the best-protected slopes and ravines along rivers in eastern Nebraska and in the southeast all of the lower slopes and the north exposures of higher ones. But both northward and westward the soil and air become drier, and at the extreme limit of its range this climax vegetation covers only the foot of the north slopes. It extends back from the river about ten miles in the south but only about one mile in the north. Along the tributaries of the lower Missouri it extends westward a distance of about 20 miles. The red oak extends up the Missouri until the river turns westward, but the linden finds suitable habitats at the foot of the bluffs and occurs along the Niobrara River as far west as about central Cherry County. Along the Platte and Big Nemaha the farthest westward extension of red oak is about 40 miles. Linden is found farther west on these streams and along both the Big and Little Blue Rivers and their larger tributaries.

Red oak grows farther north than most other oaks in Nebraska and is exceeded in its western range only by the bur oak. In the Ohio River Valley it attains a height of 100 feet or more and a trunk diameter of 3.5 feet. But in the best sites in Nebraska larger trees attain heights of 75 feet and in its northern range only about 28 feet. Diameter of trunk decreases from 24 to 8 inches. The grayish brown bark, tinged with red, is usually an inch thick on older trees.

Linden, in the center of its range, is nearly 100 feet tall and has a trunk diameter of 3 to 4 feet. The larger trees in Nebraska are about 80 feet in height and 2.5 feet in diameter in the southeast. They are only about 22 feet tall and 8 inches thick where the Missouri River first touches Nebraska. A broad rounded crown is developed in open stands, but linden adapts itself to less space in close stands and becomes taller and more columnar.

Ironwood, with no reduction in size, is closely associated with the linden. It ranges farther west than the linden along the Niobrara and extends into the Black Hills. Some trees are 30 feet high and 10 inches in diameter, but its average height in Nebraska is only about 18 feet. Among other trees occasionally found in this community, but only in the southeast, are Ohio buckeye, black walnut, Kentucky coffee-tree, and redbud. But even these are excluded where the forest cover is dense. Other examples are the pawpaw and yellow or chestnut oak. In the dense shade of this forest, shrubs are few. Prickly ash, Virginia creeper, and honeysuckle are examples.

Black Oak–Shellbark Hickory Community

A second forest community dominated by black oak and shellbark hickory occurs in places slightly less mesophytic than the preceding. This community is more limited in extent. It does not occur farther north than Omaha and its western distribution is a little less than that of the red oak. The typical community consists of trees about a foot in diameter and mostly 35 to 50 feet high (Figs. 14, 15).

Shellbark hickory is a tall forest tree with a straight columnar trunk. The bark exfoliates in long, hard, plate-like strips which

Fig. 14.—Interior view of a forest of yellow oak and shellbark hickory in extreme southeast Nebraska. The shade is dense and Mayapple is the most abundant herb on the forest floor.

Fig. 15.—Typical stand of shellbark hickory on a gentle slope near the top of a hill.

hang attached at the upper end and give the tree the appearance of shagginess. Hence it is also called shagbark hickory. Height of mature trees varies from 35 to 70 feet and diameters up to 12 inches. There is a tendency where the slopes are quite gentle and extensive for the two dominants to be separated, shellbark hickory on the upper drier part of the slope and black oak on the lower part. Cutting has been very extensive and few good examples of the forest type remained in 1928. Other trees and, in open stands, shrubs are intermixed. The most important subdominant trees are yellow oak, Kentucky coffee-tree, linden, and red oak. There is no continuous shrub layer, and all of these shrubs also occur in the much more widely ranging forest of bur oak-bitternut hickory, to be described.

From the notes of two prominent Nebraska botanists who examined these forests about 1898, we have the following. The layer of small trees and shrubs is never well developed. In the heart of these forests the species of this layer are few and the individuals of infrequent occurrence. The pawpaw, wahoo, Indian cherry, bladdernut and rough-leaf dogwood comprise practically all the species represented in this layer. With the exception of the wahoo and the rough-leaf dogwood, which form small clumps, all are solitary or nearly so. As would be expected, however, the layer finds greater expression in the forest's edge, where the forest begins to lose somewhat of its closed character. Here are to be found a host of small trees and large shrubs, which constitute immense thickets. The most common examples of these are thickets of plum, of choke cherry, of service berry, of sumac, of hazel, and of elderberry. Other species are of almost constant occurrence throughout. These are gooseberries, Indian currants and red and black raspberries. The catbrier or smilax is the only climber that dwells in the deeper portion of the woods. It is a high climber and its foliage often occupies the same level as the foliage mass of the trees. The other woody climbers, Virginia creeper, wild grape, and poison ivy, are usually found in the forest's edge, or in thickets, where they climb over shrubs and trees alike, forming a dense wall of foliage. Wild cucumber and virgin's bower are most conspicuous in thickets where they spread widely over the shrubs.

Bur Oak–Bitternut Hickory Community

This community occurs on the drier slopes and tops of hills (Figs. 1, 16, 17). The bur oak also grows along the larger streams, often far westward beyond the range of the map (Fig. 18). Here the original location of this community is shown as nearly as possible and no account is taken of disturbances which have been made by cutting for timber or clearing the land for farming or where conditions were such as to favor other types of woodland.

The bur oak community is nearly always bordered by a more or less continuous community of various shrubs which separates it from the grassland. Northward beyond the mouth of the Platte River the more mesic forest communities almost disappear. There is a decrease in the area occupied by woodland, decrease in number of species, dwarfing of individuals, and trees are confined to the most favorable sites. The area occupied by shrubs, however, is greatly increased.

In Nebraska the bur oak is a sturdy tree with large, stout branches and somewhat sparse branchlets. The bark on large trees is about 1.5 inches thick, grayish-brown and deeply furrowed. The large acorns are produced in abundance. Squirrels and other ani-

FIG. 16.—Yellow oak and black oak near Rulo in southeastern Nebraska.

mals have carried and stored them for food so widely that the tree is distributed along almost all of the larger streams. This oak is well adapted to upland sites. The taproot is usually about 9 inches deep before the first leaves are unfolded. During the first season of growth the well branched taproots may penetrate to a depth of 5 feet. Taproots of mature trees often reach a depth of 13 feet or more and the numerous, strong, lateral branches may run outward 15 to 30 feet. This tree often grows in drier soil than that required by most forest trees.

In the better protected localities of Nebraska, bur oak attains a height of 85 feet and a diameter of 2.5 feet and may reach an age of 100 years. Conversely, it may survive in places which seem too dry for tree growth. Here it is making a last stand against the rigors of a prairie climate. For example, on the upland near Norfolk there was a shallow ravine filled with scrubby bur oak. Trees from 2 to 7 feet high ranged from 20 to 28 years of age. They were only 1.5 to 3.5 inches in diameter. An average of 12 years was required for an inch of growth in width. This illustrates the ability of bur oak to grow under extremely adverse conditions (Fig. 19). In southeast Nebraska an increase of an inch in trunk diameter may be attained

Fig. 17.—Bur oak near Omaha. The trees have been thinned to promote growth of bluegrass for pasture, but there is no reproduction.

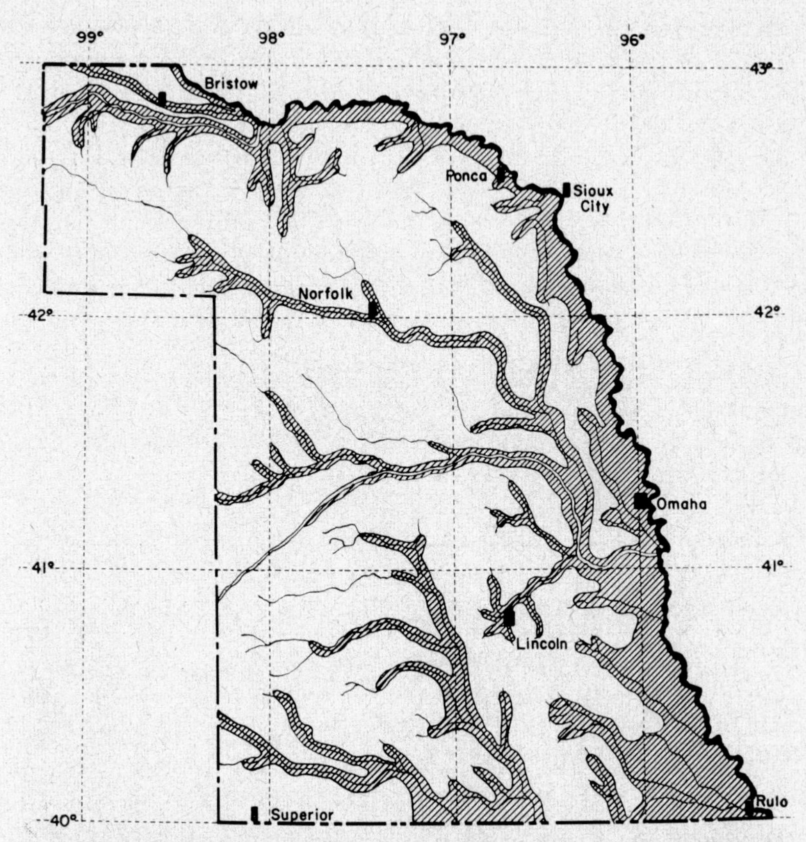

FIG. 18.—Distribution of bur oak along rivers and streams in eastern Nebraska. Redrawn from Dr. J. M. Aikman, 1929.

in less than three years. Scrubby bur oaks are common in ravines near the source of many small streams.

Bitternut hickory is also a good pioneer tree because of its manner of distribution. It has a thin-shelled nut, which, although bitter, is used as food by several species of rodents. Because its tolerance to unfavorable conditions is less than bur oak, it has a much smaller range. Moreover, a fungus disease, which caused the death of the trees after they became fully established and often fully grown, was rampant in Nebraska in the 1920's. This hickory does not seem able to reduce its size and live at the border of its range to the same degree as bur oak.

The typical bur oak-bitternut hickory community is composed of a somewhat sparse stand of trees, the oak outnumbering the hickory about three to one. The height of the hickory is about 35 feet but varies with the habitat from 20 to 70 feet. The diameter of the bur oak is 6 to 14 inches, that of the bitternut hickory 4 to 10 inches. Beneath the forest crown there is much more light than in either of the preceding communities and consequently much more undergrowth. The yellow or chestnut-oak is a secondary species which is most nearly like the bur oak and hickory in its habitat requirements. Scattered trees from other communities also occur, the remainder of the oak-hickory forest is composed of flood-plain species such as elm, ash, and hackberry, which have migrated to the upland. Conversely, the bur oak is found on the flood plains of smaller streams.

About 40 species of shrubs occur as an understory in this oak-hickory forest in the southeast. Often they are dwarfed because of the shade but they flourish in more open places. Some extend far beyond the forest border. Associated with the trees and occurring in the more dense area of shrubs are poison ivy, grape, bittersweet, and other vines. In northern Nebraska the kinds of shrubs are reduced to nearly one-half. But the area occupied by them, with a great decrease in forest growth, is much greater. Dr. J. M. Aikman has made the most extensive studies in these forests, and this report includes many of his findings.

The herbaceous flora of the forest is very different from that of prairie. Forbs of woodland seldom extend beyond the border of shrubs and then only during wet years. Not only forest trees, but also herbs of the forest, fail to appear on the open prairie. There is

FIG. 19.—Sections of 20-year-old bur oak trees from the foot (left), midway up, and top (right) of a slope in southeastern Nebraska. Diameters are 4.25, 3 and 2 inches, respectively.

FIG. 20.—Blue phlox in spring, from damp woods in eastern Nebraska.

indeed very little mingling of the two floras, which are strikingly different. In our state more than 50 species of herbs contribute to the woodland flora. None of these are found in the prairie. Grasses are very few; bottle brush and reed grass are the most common. Blossoming of many species occurs in spring before the foliage of the trees greatly obscures the light (Figs. 20, 21). Others blossom and set seed in autumn, often in October. A list of some of the more common species follows.

Bedstraw	Jack-in-the-pulpit
Beggar-ticks	Mayapple
Blood-root	Monkey flower
Blue larkspur	Nettle
Blue phlox	Orchids
Clear-weed	Solomon's seal
Columbine	Spring beauty
Culver's-root	Spring lily
Dutchman's breeches	Squirrel corn
False dragonhead	Sweet cicely
False Solomon's seal	Tall bellflower
Great bellflower	Three-seeded mercury
Horsemint	Touch-me-not

Violets, blue and yellow	Woods nettle
White avens	Woods vervain
White snakeroot	

There are also a few climbers, such as ground nut and hog peanut, and several species of ferns.

It has been estimated that 15 percent of Iowa, owing to its many large rivers and network of streams, was forested. But only 2.5 to 3 percent of the area of Kansas and Nebraska was native forest.

Shrub Community

This community in the east is found in places not occupied by the preceding woodland communities. These are the more xeric hilltops and southwest slopes. Various trees from forest communities often occur, such as red cedar, crabapple, mulberry, hawthorne, redbud, honey locust and others. The outer edge of the forest is in close contact with various shrubs, especially hazel, coralberry, smooth sumac, prickly ash, June berry, wild red raspberry, and gooseberry. Buckthorn, bladdernut, elder and several species of dog-

Fig. 21.—Woods violet is common in woods and thickets, especially along streams, over most of the state.

FIG. 22.—Young bur-oak forest at Weeping Water with a dense undergrowth of herbs and shrubs.

wood are also common. Several vines occur; greenbrier or smilax, Virginia creeper or woodbine, poison ivy, bittersweet and grape are examples. Small trees are represented by wild plum, choke cherry, and June berry.

Shrubs not only form a layer in the bur oak-bitternut hickory community but also extend beyond the forest edge into prairie (Figs. 22, 23). They thus form a zone of transition from forest to prairie. This has a width varying from a few yards to half a mile or more. That the shrub community has a greater number of species in the southern part of the state but is more extensive in the northeast results from a drier climate northward and less competition with the trees for favorable sites. Where the conditions in grassland are altered so as to considerably increase the water content of soil and decrease evaporation, invasion by shrubs is likely to occur.

Fig. 23.—Red oak and linden on a north-facing slope near Weeping Water. The shade is dense and few forbs bloom except in spring and late autumn.

FIG. 24.—Belmont prairie near Lincoln with cattails and prairie cordgrass in the ravine and sumac and choke cherry on the lower slope. Photo 1916. Such situations were common early in 1900.

Extensive measurements of habitat conditions in widely scattered shrub communities and adjacent areas of prairie have been made. Shrub habitats always showed a higher percentage of soil moisture, higher humidity, lower (summer) temperature, decreased wind velocity, reduced evaporation, and decreased light intensity. Where forests invade prairie it is almost always under a cover of shrubs. The shrubs migrate into adjoining grassland by means of rhizomes or by shallowly placed horizontal roots, as in smooth sumac, from which new stems arise. Beneath them there may slowly develop conditions favorable for the growth of trees. Seedlings of trees have little chance for survival under the dense shade of grasses.

The three most important and abundant species of shrubs are hazel, wolfberry, and smooth sumac. Their ability to pioneer outside the forest varies greatly, sumac which withstands forest shade least is the best pioneer in the grassland and is widely distributed throughout the state (Figs. 24, 25). In the northeastern part of Nebraska the entire summits of bluffs bordering forested areas along the Missouri are often covered with sumac. It reduces the light for grasses very materially. The migration of shrubs into grassland proceeds from woodland outward along gullies many of which extend far into the prairie. The more xeric ones are filled with sumac and often contain wolfberry. Indeed this species or buck brush, especially southward, may occupy extensive areas at the expense of sumac. This shrub retains its hold; it can withstand shade much better than sumac and is very common in the Bur Oak-Hickory forest. There are three species, the one with purplish fruits occurs in the woods in the eastern part of the state. It has several names, coralberry, Indian currant, and buck brush. The wolfberry, with white fruits, is common throughout the state; while the low snowberry (also with white fruits) occurs in dry, rocky soil mostly in the western part of the state. The most common one, which spreads by rhizomes into great circular patches especially in native pastures, is the wolfberry. It is a low-growing shrub which makes a dense stand (Figs. 26, 27).

Figure 28 shows a south-facing slope of bluffs which border the Missouri River in northern Nebraska. The opposite, north-facing slope is forested mostly with bur oak. This side is covered with patches of rough-leaf dogwood, coralberry and smooth sumac, with bur oak in the ravines. Figure 29 (center) shows bluffs bordering the Missouri in east-central Nebraska. Their eastern side is occupied by a forest of linden and red oak, the top by one of bur oak. Alfalfa grows on the flood plain. Figure 29 (lower) shows the west side of the same bluffs and the development of forest and shrubs over the second and third rows of hills back from the river.

FIG. 25.—Sumac along a small stream invading lowland prairie by means of offshoots from long horizontal roots. It is held in check by annual mowing.

Fig. 26.—Patches of wolfberry in the original prairie sod after 60 years of moderate grazing. This shrub propagates mostly by rhizomes.

Fig. 27.—Thicket of wild plum in full bloom in early spring. The prairie grasses, mowed in fall, are kept out by the shade.

FIG. 28.—South-facing slope in the northeast (upper) with patches of shrubs.
FIG. 29.—Bluffs (center) bordering the Missouri River in east-central Nebraska, and view of forest and shrubs (lower) on west hillsides. Explanation in text.

The rough-leaf dogwood is able to grow among the sumac and the developing bur oaks. It follows the sumac in its advance into grassland but gives way in the shade of the bur oak after the trees attain a height of 10 to 15 feet. Hazel usually occurs in patches near the forest edge and as an understory within it. It is least xeric but can shade out the preceding shrubs. It spreads into Nebraska grassland only at a very slow rate.

It has been conclusively shown that trees cannot successfully invade undisturbed true prairie. Conditions on low prairies are unfavorable for tree and shrub seedlings because of the dense shade cast by the tall grasses, such as big bluestem. On uplands, soil moisture is insufficient to sustain their growth in competition with the grasses under the prevailing dry grassland climate. This is not the opinion of a layman but the conclusion after long-time experimentation and observation.

The wild plum is a small tree which frequently forms shrub-like thickets in ravines in grassland throughout the state. A few western or northern shrubs do not occur in the southeast; examples are buffalo berry, western June berry, mountain mahogany, and western choke cherry.

III.

Flood-Plain Forest

Forest communities disappear rapidly along streams westward from the Missouri River and the typical condition is a mingling of upland trees with those of the flood plain. Farther from the great river, trees are mostly confined to the banks of larger streams and to broad shallow ravines tributary to them, or to the vicinity of springs. Some are found in the shelter of steep protecting bluffs. In the driest grassland trees occur naturally, if at all, only as small groups or as individuals in the most protected places. The valley type or flood-plain forest is represented along nearly all of the streams. It is the common woodland type in Nebraska. Except for small amounts of red cedar it is composed wholly of broad-leaf species. In the western part of the state it is confined closely to the water courses. Broad-leaf species require more moisture than conifers and in the west no good growth of them is found in any place where the roots can not reach nearly to permanent water.

Westward especially, the trees are low with spreading crowns, though dense stands sometimes occur in moist situations which produce good clear trunks and attain a height of 75 feet. Otherwise the height of broad-leaf species is usually less than 40 to 50 feet while many mature trees do not reach this height and have diameters of only six inches or less. Only a few species of woody plants occur along the streams of central and western Nebraska, and these are noticeably smaller in every way than those in the eastern portion of the state. A height of 20 to 35 feet is common, and the deciduous trees are often confined to the lower banks of the streams. The lower branches usually occur only 5 to 10 feet above the soil, depending largely upon the degree of protection from winds. The drought-enduring deciduous species that grow in western Nebraska are even fewer and smaller. They are restricted to some of the most protected sites and often occur as individuals rather than in groups.

The Republican and Platte Rivers are preeminently sandy streams with shifting beds so that timber growth is either wholly absent from them over long stretches or consists only of scattering cottonwood and willows. Smaller streams which are less sandy support other species, principally green ash, white elm, hackberry and boxelder. Cottonwood and willow are the only native broad-leaf trees which seem to prefer a sandy soil and tolerate a high water table so they often form the sole tree growth along the larger streams. Cottonwood stands are generally open. Each tree has but little modifying contact with its neighbors. On minor streams which have the heavier soil other species dominate and cottonwood and willow are less abundant.

As one travels westward in Nebraska there is a progressive decrease in the number of tree species until the foot-hill region is reached. Here the increased altitude results in the appearance of certain Rocky Mountain forms. In Buffalo County, for example, the species found are bur oak, green ash, boxelder, white elm, hackberry, cottonwood, and willow. Farther west in Deuel County the boxelder, elm, and oak are missing, while in Scotts Bluff County occur boxelder, several willows, cottonwood, green ash, hackberry, mountain mahogany, and western birch.

Many streams with their headwaters in prairie have been investigated. Runoff water in a ravine begins to cut into the prairie sod to form a channel. As the channel widens and deepens it presents a habitat where wind-blown seeds of various willows and sometimes cottonwood germinate and the seedlings develop (Fig. 30). A few miles farther down the valley a small stream with banks 2 to 3 feet high may occur. It may be dry in summer or joined by spring-fed branches and cease to be intermittent. Soon the stream has wind-protected banks which are very favorable to several shrubs such as wolfberry, indigo-bush and elder. Sumac and gooseberry are commonly found and vines of grape and bittersweet, while willows and cottonwoods larger than those upstream are common. Even small trees of boxelder and green ash may occur. Thus, pioneer trees and shrubs at the streams' source are those with light, wind-blown seed. The pioneer shrubs and vines have showy edible fruits carried by birds. This early stage of woodland development is common along the small tributaries of rivers in prairie and plains.

A second stage occurs where the stream develops a flood plain with wide protecting banks. Various animals, especially timber squirrels, have gradually carried large fruits such as those of bur oak, hickory, walnut, and others from downstream woodland. Thus, as soon as a suitable habitat is provided in wind-protected places along the stream, trees and shrubs may, in the absence of prairie

Fig. 30.—First stage in the development of woodland along streams, black willow and peach-leaf willow. Explanation in text.

fires, replace prairie grasses. But on unsheltered and wind-swept banks of prairie streams, especially far from large rivers, little or no woody vegetation may occur.

A third stage is found along small streams in eastern Nebraska and in better protected situations westward. Here white elm, red elm, and green ash are the dominants (Fig. 2). Associated with these are soft maple, sycamore, hackberry, and choke cherry. Willow, cottonwood, and boxelder disappear when a forest canopy is developed. Gooseberry is perhaps the most typical flood plain shrub, but dogwood, prickly ash, wild black raspberry, and Virginia creeper are common. In the southeast black walnut becomes an important tree; Kentucky coffee-tree, Ohio buckeye, sycamore, and others also occur.

Farther down the stream an increasingly large number of both trees and shrubs occur as well as a beginning of their separation into different habitats. Usually fine large trees of red or slippery elm and white or American elm are most abundant on the banks. Principal species on the flood plain are boxelder, green ash, hackberry, honey locust, wild black cherry, and walnut. Some of these also grow at the base of slopes. A scattered growth of young bur oak may occur on protected north-facing slopes and on the sides of lateral ravines. Scattered widely over the flood plain are rough-leaf dogwood, black raspberry, and wahoo. Numerous vines occur and often climb high into the trees. Among these are greenbrier, poison ivy, Virginia creeper, and virgin's bower.

Summarizing, pioneer trees at stream sources are those with light, wind-blown seeds, such as the willow. They usually appear soon after the prairie sod is weakened by erosion. Boxelder, elms, and ash, all with windblown seeds, occur as soon as there is a favorable habitat. Pioneer shrubs and vines like elder, bittersweet, and grape have showy edible fruits carried by birds. This early stage in woodland development is represented for considerable distances along nearly all small tributaries. When a stream develops a flood plain with wide protecting banks, large fruits such as those of walnut, hazel, bur oak, etc. are carried upstream by various animals, especially timber squirrels.

The flood-plain community reaches its best development along the larger streams in the southeastern part of the state. Here black walnut takes its place as a dominant along with red and white elms and green ash. Since the shade is dense, few of the less tolerant species are found. The individual trees are much larger and in every way the flood-plain forest is better developed. Other species are the white ash, choke cherry, Kentucky coffee-tree, Ohio buckeye, hackberry, silver maple, and sycamore.

The Missouri River bottom is a vast, nearly level plain of alluvial sediments. Rivers, both large and small, as they flow through erodable land develop a flood plain. Such areas are commonly designated as bottom lands and more particularly as first or low bottoms and second or upper bottoms. Bordering Nebraska the river valley has from 40 to 50 percent first bottom land. Flood-plain forest along the river consisted very largely of willows and cottonwood. They ranged from the water's edge to a distance of ⅛ to ½ mile from the river above its juncture with the Platte and to a somewhat greater distance in places southward. Willows and cottonwoods clothed great sand bars in the river and also bordered abandoned river channels (Figs. 8, 31). Lakes and ponds sometimes covered many acres of alluvial deposits on the lower flood plain. The remainder of the low, wet, first bottom land was occupied by marshes or elsewhere by tall coarse grasses alternating or mingling with various shrubs. On first bottoms, especially within a mile or two of the stream, the river may occasionally change its course, but elsewhere such changes are to be reckoned in terms of hundreds or even thousands of years.

Fig. 31.—Sandbar in the Missouri River. From the bare center of the bar development occurs in several stages. First is sandbar willow, then black and peach-leaf willow, and, in the background, tall trees of cottonwood.

Sandbar willow is the first tree or shrub in this area to grow upon sandy or muddy banks of rivers, streams, and lake shores. Its extremely abundant, fiber-like roots enable it to maintain a hold on the soil. This small tree, usually about 20 feet high, has a slender trunk 2 to 3 inches in diameter. In dense stands, which result from its spreading by long stoloniferous roots to form thickets, it is commonly dwarfed into a shrub only 5 to 6 feet high. Branches are slender, erect, and flexible. The elongated narrow leaves are light yellow-green. Saplings 3 to 6 years old often occur at the rate of 8 to 10 per square foot. Usually there are few other plants in such dense stands. This willow is common in drying lakes, ponds, and marshes. It is often intermixed with cattails and river bulrush. Thickets of willows covering many acres are not infrequent. Sandbars in great rivers are often completely clothed with this species. Because willows stabilize the soil, the sandbars may eventually become more or less heavily wooded. At Fremont and a few other places along the Platte and on several islands in the upper Missouri in this area, red cedar also occurs on the lower flood plain and on islands in the river.

Two other species of willow are common. These are trees about 20 to 40 feet tall and 8 to 10 inches thick. They are the peach-leaf and black willow. They prefer wet lower banks of streams or borders of lakes, ponds, and marshes. They cannot withstand much shade and usually form a border to the cottonwood forest which follows the Missouri River all along its course.

Cottonwood is the most typical tree on the banks of larger rivers and often the only one on eroding shores where the fringe of willows on lower ground has been swept away. Cottonwood grows rapidly and may attain a height of 70 or more feet with trunks 3 to 6 feet in diameter. Such trees are often only 70 to 80 years old.

FIG. 32.—View on a flood plain near Weeping Water. The trees are mostly green ash, walnut, and white elm. Photo, 1929.

FIG. 33.—Large hackberry in a well-developed flood plain community on the Big Nemaha River near Falls City. It is 3 feet in diameter and 70 feet high. A white elm in the same area was 3.5 feet in diameter and 95 feet high. These, of course, are unusually large specimens. Photo 1928.

Where well lighted on forest margins or in open stands, branching begins at a height scarcely greater than that of the shrubs, but in the forest the straight, erect or more or less leaning trees are free from branches to a height of 20 to 30 feet. The bark on young trees is smooth and light yellow-green to nearly white. But on old ones it is ash-gray, thick, and deeply furrowed. Cottonwood alone does not form a dense forest. The large trees are widely spaced, often in small groups of 3 to 5 or more. It is the tangle of shrubs and vines that slows one's progress.

Cottonwood and willow often form the only tree growth along the larger streams. On the minor streams, which have the heavier soil, other trees are dominant and cottonwood and willow are less abundant. Many of the numerous shrubs already mentioned are plentiful and there is an abundance of woody vines (Figs. 32, 33).

IV.

Lakes, Marshes, and Other Wet Land

On the wide alluvial bottom land of the larger rivers, forests are mostly limited to relatively narrow strips along the channels and the abandoned oxbows, alternating with strips of grassland. This is in accord with the observations of explorers, pioneer settlers, and early land surveys, and is confirmed by studies in the early part of the present century. Along the bordering Missouri River the flood plain forest, shrubs, and coarse grasses occupied most of the first bottom. But the second bottom, which is usually about 10 or more feet higher, was nearly all covered with prairie. Exceptions occurred around ponds and lakes and on poorly drained land.

The Missouri and Platte Rivers have very wide valleys and although the bordering bluffs and hills furnish protection against the prairie climate, this does not extend throughout their width. Only a brief account will be given of the vegetation in lakes, swamps, and marshes and the coarse grasses of wet land.

Lakes and Marshes

A lake is an inland body of standing water occupying a basin; a pond is a lake of slight depth. Swamps are places where the normal summer water level is continuously above the soil surface. In Nebraska lakes occur mainly in the Sand-Hills Region and along the large river valleys. Marshes often border the lakes and ponds. It has been estimated that 320 square miles of permanent and intermittent lakes and marshes occur in Nebraska.

The chief lake region is in southern Sheridan and northern Garden Counties. Another region occupies much of southwest Cherry County, and extends southeast across Grant County and northeastern Arthur County. There are several others of smaller extent. Besides these there are many small lakes scattered throughout the sand hills and occasionally in the foothills. The lakes occupy valleys between ridges and steep sand hills. The general

FIG. 34.—Zones of white water lily, bur reed, and sandbar willow at the edge of a lake.

shape, conforming to that of the valley, is oblong. The waters are clear and shallow. The margin, which is destitute of woody vegetation, and the bed are sand.

The vegetation of lakes is composed of submerged, floating, and emerged hydrophytes. Water lilies are anchored in the soil 3 to 8 feet below the water surface by roots and rhizomes. Their large leaves often covered the water surface with floating mats of green over hundreds of acres. Pondweeds also grew in wide expanse (Figs. 34, 35, 36). In ponds the tiny free-floating duckmeats often grew so thickly that the entire water surface supported a living sheet of green. The various waterlilies, arrowheads, numerous pondweeds, and smartweeds are often the most conspicuous, but many other species occur (Figs. 37, 38).

The dominant species of swamps are all large perennial plants—great bulrush, river bulrush, broad-leaved cattail, reed, bur reed, arrowhead, and water plantain. While species in swamps are often intermixed, more frequently one species may occupy large areas, even hundreds of acres, almost to the exclusion of others. They root on the bottom of swamps and occupy definite places which are largely related to the depth of the water. The great bulrushes usually grow in the deepest water, sometimes in excess of 5 feet.

Fig. 35.—Pondweed with leaves floating in the shallow water of a lake.

Fig. 36.—Wild rice growing in a sand-hills lake.

Fig. 37.—Arrowheads growing in shallow water.

The reed grows where the water is shallowest and cattails usually at intermediate depths. All have large, much branched rhizomes and may rapidly extend their area by means of them.

The great bulrush grows early and rapidly in spring from its food-stored rhizomes. The stems are circular in cross section, and may have a width of ½ to ¾ inch and a height of 7 or more feet.

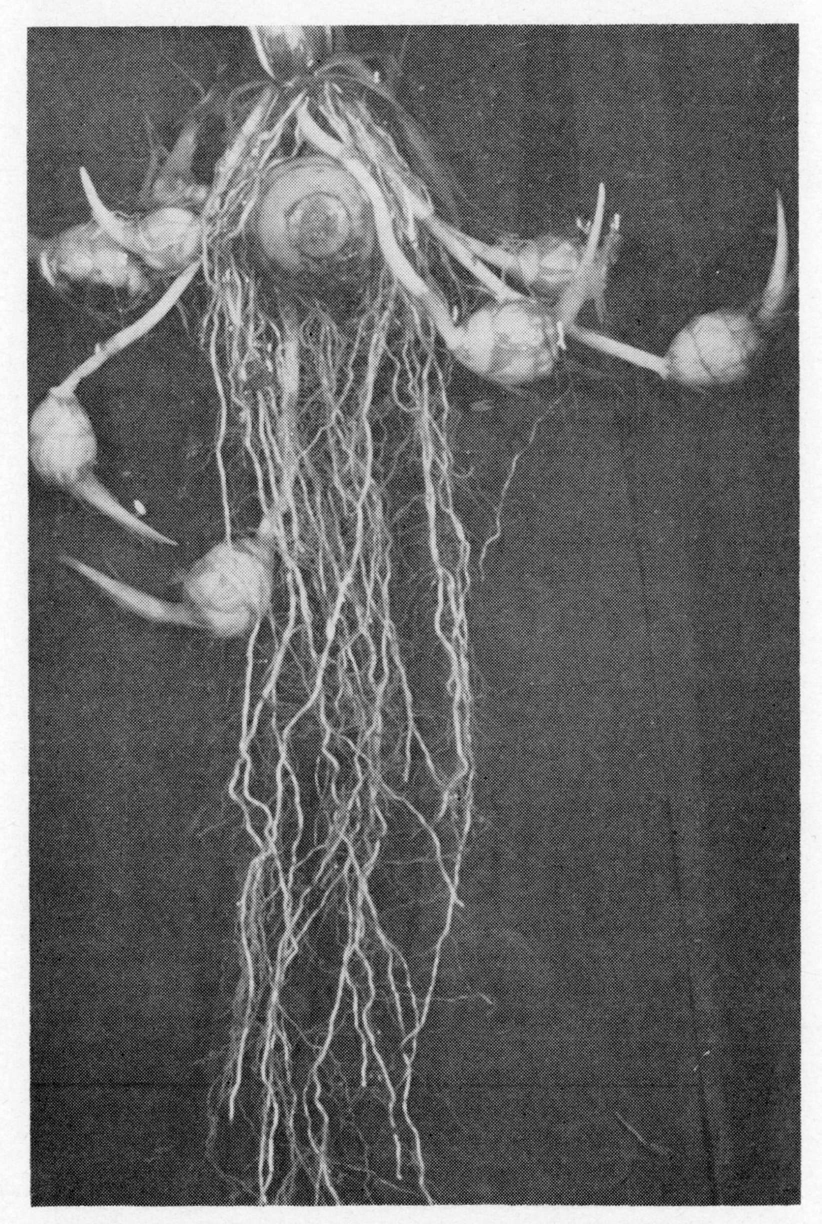

FIG. 38.—Underground parts of arrowhead—roots, rhizomes and tubers.

The flower clusters at the top of the plant are not very conspicuous. Vast areas of river banks and shallow water of swamp lands were occupied by bulrushes before the flood plains were drained. Many lakes were bordered with bulrushes continuously for several miles.

Cattails almost always occur where shallow water permanently covers the soil surface. The plants grow thickly in dense, pure stands. Examination reveals that the individuals are spaced at intervals along the large rhizomes. Height varies greatly; often it is 5 to 7 feet above the water surface. Cattails border ponds in a wide zone, and may extend along the shores of lakes as a continuous stand for miles. Propagation by rhizomes is so efficient that an entire community may develop from a few plants.

The reed is a coarse perennial grass with an intricate system of very stout rhizomes which occupy the soil to 8 or more inches in depth. It may also spread widely by means of long, horizontal stolons. It is conspicuous for its dense stands and rapid growth to a height of 8 to 12 feet. When in bloom the long hairs of the spikelets intertwine to form a silky mass.

River bulrush has coarse, stiff stems which are triangular in cross section and attain heights of 4 to 7 feet. Its numerous long leaves spread widely. Water plantain and Indian rice are other swamp plants. Indian rice, found in the sand hills, germinates under water and grows submerged for a time after which the leaves float on the water. Then erect growth occurs and the grass often reaches a height of 7 to 8 feet. A single plant may occupy an area of several square feet.

Marshes

When the soil in swamps is built up to or above the water level, marsh plants begin to colonize the area once occupied by bulrushes or cattails. Marsh vegetation is rooted in waterlogged soil where the water level in summer is close to the soil surface. Some marshes are inhabited almost entirely by tall species of grass-like sedges 2 to 3 feet high, rushes, and spike-rushes. In others the plants are mostly hydric grasses, such as reed canary grass and rice cutgrass. Smartweed, water hemlock, Canada anemone, iris, and various mints also inhabit the marshes (Figs. 39, 40). Marshes, usually of small extent in wet ravines or depressions in wet lands, are often entirely populated with water-loving herbs, mostly various smartweeds. The general appearance of marshes does not differ greatly from that of luxuriant prairie. The densely grouped sod-forming sedges are nearly all perennials which renew growth very early in spring and

FIG. 39.—Blue flag or iris from a marsh in eastern Nebraska. (From *North American Prairie*.)

are 2 to 3 feet tall in June. The total area occupied by marshes is very large.

COARSE GRASSES OF WET LAND

Between the marshes and the well drained bluestem prairies, great areas of the flood plains were clothed with various coarse grasses. Chief of these was prairie cordgrass (Fig. 41). This grass occurred over hundreds of square miles on first bottom lands along the Missouri River and its tributaries. It also grew in the edges of sluggish streams or ponds. It occupied soils too wet and too poorly aerated for the development of big bluestem. It indicates soil too wet without drainage for cropping, at least in spring, but furnishes excellent hay if cut two or three times each growing season. Yields are often 3 to 5 tons per acre. Smoothness of the leaves makes the

Fig. 40.—Fox sedge (right) nearly 3 feet high and in bloom on July 5. A rush is on the left.

hay difficult to handle; it easily slips off the hayrack or the haystack. Prairie cordgrass is taller and coarser than most grasses of lowlands. It was dominant over large areas because of its great height (6 to 10 feet) in dense pure stands, which resulted largely from propagation by rhizomes. Its shade is so dense that most other plants are excluded. Although this grass renews growth rather late, often about mid-April, it grows more rapidly than any of the prairie grasses. By the first of June it is quite leafy and about 2 to 3 feet high. Each stem has 6 to 10 leaves with rough margins and is nearly ¾ inch in width and 2.5 to 5 feet long. The flower clusters are very large. The peak of the flowering period is in mid-August. This grass is much less injured by soil deposit than are most grasses. The hard, sharp-pointed shoots may push their way upward through a foot of soil.

Switchgrass occurs in the drier portions of the prairie cordgrass community along with nodding wild-rye in soils that are too poorly aerated for big bluestem. It also occurs in ravines of uplands where it is often flanked by narrow strips of big bluestem. It is a warm season, perennial grass which renews growth late in April. Development is rapid. Early in June the foliage is usually 1.5 feet high and it completes vegetative growth in mid-July. Flowering then begins, reaches a maximum in August, but continues until late autumn. The stems are widely spread along the rhizomes. The panicles are very large, 10 to 20 inches long and 16 to 20 inches wide (Fig. 42). Switchgrass is readily eaten by livestock. After the stems become woody, the leaves and panicles are eaten. It should be cut for hay twice each summer. It produces a high yield.

FIG. 41.—Prairie cordgrass about 6 feet high and rhizomes and roots in the upper soil.

FIG. 42.—Typical stand of switchgrass 3.5 feet tall on June 24. The old flower stalks of the previous summer are 5 to 5.5 feet tall. Reed canary grass (right).

Reed canary grass is a coarse perennial which grows in wet places in soil too poorly aerated for most other grasses. Like most species of low ground it renews growth early from a vast system of coarse, deep, tangled rhizomes. Mature plants range from 3 to 8 feet in height.

Nodding wild-rye is often associated with switchgrass and thrives best under a condition of soil moisture intermediate between those of prairie cordgrass and big bluestem. It is more abundant northward than elsewhere. This grass reaches a height of 3 to 4 feet by mid-June. Then the flower spikes begin to appear and a total height of about 5 feet is attained (Fig. 43).

Eastern gama grass, found only in wet places in the southeastern part of the state, is a tall perennial plant which grows in habitats similar to those of prairie cordgrass. It may also be found in seepage places on hillsides. Its large circular bunches range from 1 to 7 feet in diameter. This grass renews growth in March from a mass of extremely large and compact rhizomes. The massive foliage is 3 to 4 feet high by mid-July. Height of the branched flower stalks is 5 to 7 feet. The somewhat woody stems are flattened and not circular like those of most grasses. A very coarse type of hay is produced in great abundance. A single cutting may yield 2.5 to 3 tons per acre and often three cuttings are made during the growing season.

Among other grasses of wet lands are rice cut-grass, which grows in shallow water, American sloughgrass, bluejoint, northern reedgrass, redtop, fowl mannagrass, and many others.

Thousands of acres of first bottom land along the Missouri River and its tributaries were covered with marshes, prairie cordgrass, and wet meadows of switchgrass and other coarse grasses. All were subjected to flooding and soil deposit. Numerous forbs and flood plain

FIG. 43.—Indian grass and Canada or nodding wild-rye.

shrubs were more or less intermixed. Many of these will be mentioned in the lowland prairie next to be described.

This study, begun in 1916, has endeavored to picture the native vegetation in its original condition as a scientific record of the past. Since the turn of the century drainage districts have been formed. Later, powerful machinery was used in clearing away trees and digging deep drainage ditches. Native vegetation throughout the bottom lands almost everywhere, except near the river channel, has been almost completely replaced by farm crops.

Depressed Areas

Scattered throughout the nearly level prairies were depressed areas, commonly known as buffalo wallows. These were occupied by a different vegetation than the seemingly endless carpet of grass. They varied greatly in size. Many were small, covering only one-fourth acre; others were 80 to 160 acres in extent. The largest were sometimes 1 to 3 miles long and 2 to 4 square miles in area. The smaller depressions were only a foot or two below the general soil level but the larger ones were depressed 10 to 15 feet. Depth of accumulated water varied greatly from year to year and from spring to autumn. The shallow depressions usually became dry by mid or late summer. In larger and deeper ones water exceeded 3 feet in depth during wet seasons. These fresh water marshes were scattered thickly over the plain. It has been ascertained that they occupied an area of more than a hundred square miles.

Soils with heavy and thick clay pans have developed in these depressions. The water seeps through the heavy soil, or more probably through great cracks formed by the shrinking of the clay; hence there is no accumulation of salts. In these places, which are more or less alternately wet and dry, only adapted species of plants can exist.

One of the more abundant species and one which gave the characteristic color tone to these depressions was a tall bulrush. A dark blue to green or even brownish color prevailed where dense stands occurred in places where the water was more or less continuously deep. In shallower water, perhaps only 18 inches deep in spring but nearly absent in late summer, a spike-rush grew thickly in tufts 12 to 18 inches high. This perennial gave the landscape a very dark blue-green appearance throughout the growing season. In less permanently wet zones emersed species of smartweeds, various pondweeds, and water plantain were common as were also cattails, big bulrush, and spike-rush.

Travelers and settlers were much impressed with these places, which were sometimes explained as "buffalo wallows" but were probably not that at all. As a result of close grazing and cropping such places have largely disappeared.

PRAIRIE PLAINS

The Prairie Plains (Fig. 7) is one of the leading hay producing districts in the United States. Most of the land is too wet for cultivation. The production of hay is confined largely to the Elkhorn Valley and the sand-hill areas of Rock, Brown, and eastern Cherry Counties (Figs. 44, 45). The soil is sandy over almost the entire area. Most of the rainfall, about 22 inches annually, enters the ground as rapidly as it falls. There is only a meager system of creeks and rivers, and only a few of the hundreds of valleys are drained by surface drainageways. Many valleys are drained by an underground system; some of the water appears in marshes, lakes, and streams but most of it remains underground over long periods of time. The topography of this valley area is relatively uniform, but knolls and even sand hills sometimes occur.

The ground water lies near the surface in most of the Elkhorn Valley region and in many of the sand-hill valleys. It varies within a range of about three feet throughout the year and is highest in spring and early summer. An intricate relationship exists between the depth of the ground-water level and the kind of native vegetation.

FIG. 44.—Stacks of hay near Ewing along the Elkhorn River.

Fig. 45.—Flood plain of the Platte River near Fremont, showing trees along the river in the background and the wide expanse of grassland. Photo July 15, 1929.

There are four plant communities, aside from typical sand-hill vegetation. There are wet areas of the prairie cordgrass type; areas of big bluestem and Indian grass; drier areas of the little bluestem and Indian grass; and the dry mixed prairie of the needlegrass and grama grass type. For example, in a representative area in mid-July, reed canary grass, prairie cordgrass, and a water smartweed composed much of the vegetation where the water table was only 44 inches deep. Redtop, switchgrass, and Indian grass were plentiful where the water table was 53 inches deep. But where its depth was 65 inches big bluestem composed the bulk of the vegetation but little bluestem was also present. Needlegrass and blue grama often pervailed where the water table was 20 or more inches deeper than that under the bluestems. A meadow of big bluestem yields one to two tons of hay per acre; the wetter types yield more forage but it is of inferior quality; the drier types yield much less than the bluestems.

V.

Grasses of Bluestem Lowlands

Big bluestem is one of the two most abundant and controlling grasses of the prairie. It and little bluestem constitute fully 75 or more percent of the true prairie in Nebraska (Figs. 46, 47). The taller grass is more mesophytic than the mid grass and is best developed on lower slopes, which receive run-off water, and on well aerated soils of lowlands. Here it composes 80 to 90 percent of the grassland. Little bluestem forms the greater part of upland prairie and in amount also exceeds all other upland grasses combined (Fig. 48). Lowlands and uplands are used here as relative terms to indicate soils of higher or lower water content, respectively. On lower and mid slopes the bluestems intermingle, often in about equal amounts. Big bluestem also occurs in amounts of 5 to 25 percent except on the driest hilltops. Thus, there is a good reason for the name bluestem prairies.

A seedling of big bluestem develops new shoots or tillers at the age of seven to eight weeks. In fact the tillering habit is pronounced. Rhizomes or underground stems are also produced abundantly. They spread widely and become much branched. Thus, the seedling increases in area both above and below ground. Moreover, this grass is very tolerant of shade and develops in much shaded places. Big bluestem associates mainly with other tall grasses common on well moist soils. Any grass which attains a height of 5 to 8 feet belongs to this group. Conversely, little bluestem, like most upland grasses, is of medium height—2 to 4 feet—and is a representative of the mid grass group. Short grasses of the western plains are of lesser height, 0.5 to 1.5 feet. Thus, in competition with big bluestem for light, little bluestem is often shaded out. On mid slopes and dry lower hillsides it intermingles with big bluestem on more or less equal terms. Seedlings soon attain a height of 6 to 8 inches and tiller profusely to form compact bunches, which in a few years attain a basal diameter of 3 or more inches.

The coarse fibrous roots of big bluestem and many other lowland

FIG. 46.—New growth of big bluestem showing the sod habit .

grasses are 6 to 8 feet deep. Their abundant foliage is held high above the soil and makes heavy demands for a supply of water. The characteristic habit of growth, the relation of roots to tops, and certain other features of the three great dominants of lowland prairie are shown in Figure 49. The great foliage height of big bluestem, about 3.5 feet, is exceeded by that of switchgrass, and both are surpassed by prairie cordgrass. Flower stalks of switchgrass may attain a height of 6 to 7 feet and those of big bluestem and prairie cordgrass 6 to 10 feet. Only the amount of tops that occurred in a strip of soil one inch in width is shown in the figure. Likewise the average number of roots is from an inch-wide volume of soil. These data were obtained for each foot in depth from perpendicular walls of trenches from several different soils. Relative width of the main roots of the three grasses is shown, but only enough detail is added to indicate the branching habit.

The fine fibrous roots of little bluestem and other dominant upland grasses (shown later) are mostly about 5 feet deep. Moreover, the coarse roots of the tall, sod-forming grasses of lowlands grow directly downward from the rather continuous mat of rhizomes while on uplands many roots spread widely, often 2 feet from the bunches, in the upper soil. Even big bluestem when growing in bunches on uplands modifies its root habit until it is similar to that of other upland grasses. Both bluestems are of southern origin; they are warm-season grasses and renew growth late in spring, usually about mid-April. Like nearly all prairie grasses they are not only perennial but have a long life span; it is believed that they live 15 or more years.

Dominants are the most important or controlling plants. They

FIG. 47.—A good stand of big bluestem on a flood plain in September. (From *North American Prairie*.)

Fig. 48.—Little bluestem prairie on upland in midsummer, illustrating the bunch habit. White prairie clover overtops the 18-inch-high grass on the left; lead plant occurs in the center and a blazing star on the right. Prairie cat's-foot grows near the soil surface. Photo 1933. (From *North American Prairie.*)

are well adapted to the environment or place in which they live; they are perennial, which insures permanent occupation; and they attain large size compared with other competitors in the upper or lower layer of vegetation, respectively, where they spread their foliage. The dominants of true prairie are those best adapted to a prairie climate.

Development of big bluestem is rapid in spring. By July 1 it is 1 to 1.5 feet tall on upland and 1.5 to 3 feet on lowland. Height attained by flower stalks varies from 3 to 5 feet on uplands but 8 to 10 feet on well watered lowlands. They begin to appear in midsummer and are abundant in autumn (Fig. 47). The beautiful autumnal colors on upland are mostly due to the wonderful shades of red, yellow, and bronze afforded by the drying plants of little bluestem, just as those on lowland are largely due to the foliage of big bluestem. Big bluestem is the best prairie grass for pasture or hay. It is highly palatable and nutritious and very productive; cattle and horses show a great preference for this grass. It makes excellent hay, especially if cut in early bloom before the stems become too hard and fibrous for palatability. Two cuttings each year are often

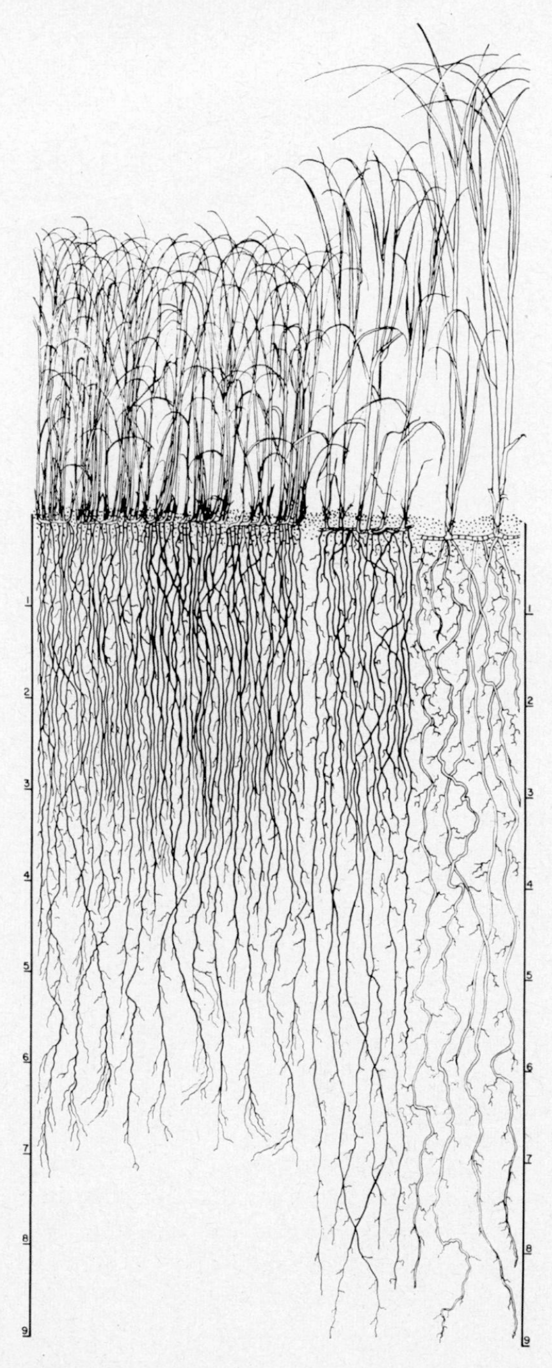

FIG. 49.—Characteristic development of tops and roots of big bluestem (left), switchgrass, and prairie cordgrass (right). When flower stalks are fully developed and flowering occurs, heights of 6 to 10 feet are attained.

made on lowlands. Yields of 2 tons per acre are common and sometimes they may exceed 3 tons.

Little bluestem often forms sod mats which consist of small tufts closely aggregated. But usually the bunches are spaced quite apart, and the spread of foliage is at least twice that of the base of the bunch. This grass is highly palatable throughout the true prairie, especially during early stages of development. With big bluestem it ranks high in the great permanent prairies of the Kansas Flint Hills and in the Osage Hills of Oklahoma. Here they will remain, unless grazed out, since the soil over the limestone rock is too shallow for cultivation, though the roots of prairie species penetrate deeply here as elsewhere. Properly managed native pastures are dominated by bluestems. If harvested early, little bluestem produces an abundant crop of good quality hay.

Warm season grasses, as illustrated by the bluestems, side-oats grama, and buffalo grass, have come from a warmer, southern climate. They renew activity much later in spring than cool-season species. They produce much foliage in midsummer. Flowering and seed production occurs from midsummer to late autumn and varies somewhat with the species. There is little late fall growth.

Cool-season grasses renew growth early in spring and make their maximum development from late March to early June. They reach maturity and produce seed in late spring and early summer. During hot weather they are more or less semidormant, but growth is usually resumed during the cool months of autumn. Some remain green despite frosts and thus greatly extend the period of green forage for all grazing animals. They are of northern origin. Needlegrass and western wheatgrass are examples as is also Kentucky bluegrass.

Monolith Method

A new method was devised in 1948 by which a sample of an entire root system from soil surface to maximum depth of penetration may be taken. It can be separated from the soil without injury to the root or displacement of individual roots from their natural position, and examined in the laboratory in relation to the various horizons of the soil profile. The monolith method is of great value in comparing root development and activities of roots in various soil types and at all soil levels. Since it has been used by various researchers in studies on wheat, corn, alfalfa, and sweet clover, we will digress here to explain how the roots in Figure 50 were obtained.

A trench about 3 feet wide and 4–5 feet long is dug in a site

Fig. 50.—Roots of big bluestem 6 feet deep (left). Those of little bluestem (center) and buffalo grass (right) are from soil monoliths 4 feet deep. Each sample is 12 inches wide and from a monolith of soil 3 inches thick. (From *North American Prairie.*)

where there is normal development of vegetation. The depth is usually 4–6 feet. Beneath the particular sample of grass, previously selected and left undisturbed in the side wall, the wall of the trench is made smooth and vertical, as shown by a plumb line. A long, shallow, wooden box, 12 inches wide and 3 inches deep (inside dimensions), without a top and lacking one end, is employed. It is placed on end, with the closed end downward. The open top is placed against the vertical trench wall, the upper end of the box just reaching the soil surface. An impression of the sides and lower end of the box is made on the wall of the trench by tapping the bottom of the box vigorously with a four-pound sledge hammer. The box is then removed and the soil column marked out with butchers' knives having rigid blades. The soil on the sides and below these marks is removed by means of knives and spades until the monolith protrudes from the trench wall, its sides and bottom extending outward at least 3 inches. The box is then fitted tightly over the monolith and the bottom and lower end of the box are braced to hold the soil column in place. Finally, the soil on the inner, attached face of the monolith is cut away by working inward with knives and spades from each side. The soil is not cut close to the top of the box, but a v-shaped ridge of soil is left protruding throughout its length. This is a part of the sample when the braces are removed and the monolith is lifted out of the trench. Then the monolith is reduced to exactly 3 inches in thickness by removing the ridge of soil.

The soil is removed from the box by a process of repeated soakings, often for several days, and gentle washing, mostly under water, even when it is extremely compact or contains a claypan. A flaring rose nozzle attached to a garden hose is employed. During this process one may observe the intimate relations of soil and roots. Roots are unharmed and in their natural position in the water after the soil has been washed away. Efficiency and success are gained only by experience. It is a long process, but one feels well repaid for the time spent when, after three to five hours, the root system alone is left in almost perfect condition in the bottom of the box. The root system is transferred to a background of black felt and photographed.

The great detail of root branching that is revealed by this method is illustrated by western wheatgrass. A monolith from a claypan soil where wheatgrass grew in a pure stand was taken on a hillside near Lincoln. The roots had three very different habitats. The A horizon of this soil consisted of a black, well granulated silty clay loam 12 inches deep. This gave way abruptly to a blocky, prismatic structure, B horizon, which extended to a depth of 28 inches. The clay content increased rapidly with depth and this hard claypan horizon was removed only with great difficulty. This dark brown subsoil gave way to yellowish loess of the C horizon, which is a silty clay loam with massive structure and much smaller clay content. In this yellowish parent material lime was abundant. The soil was easily removed and roots were readily separated. The soil broke into blocky pieces and most roots were compressed and flattened between the platy lumps. This condition maintained to a depth of 4 feet.

Cross sections of the 3 x 12 inch root sample from each soil horizon are shown in Figure 51. There were about 515 roots in the A horizon; 165 extended into the B horizon but they decreased to 85 at 2 feet in depth. The dry weight of roots in the B horizon was less than a third of that in the layer above. The C horizon had a root distribution distinctly different from that in the A and B. The roots were often flattened on the faces of the small blocks which cleaved in all directions. Thus branching occurred in all planes. Hence, when the soil was gently washed away, there remained a glistening white mass of material with branches running outward at all angles. The root habits in the three soil horizons were as different as the environment each presented. Ordinarily wheatgrass is quite uniform in its root distribution, the amount of roots gradually diminishing with depth to 4–6 feet.

Rhizomes

Rhizomes of most prairie plants, whether grasses, sedges, or forbs, are relatively shallow. They are mostly confined to the upper 4 or 5 inches of soil. They not only propagate the plant but they are also a storehouse for food. Some may live many years. This part of the plant is well protected from extremes of heat from prairie fires, ravages by grasshoppers, or long continued grazing and trampling. Rhizomes of grasses may readily be distinguished from roots by the presence of nodes or joints to which scale-leaves are attached. The rhizome ends in a terminal bud, which, like other buds distributed along its length, may either develop new shoots or produce new rhizomes.

The process of translocation and storage of reserve materials in underground organs of grasses has been summarized by a plant chemist as follows. Certain carbohydrates (mainly sugars, fructosans, dextrins, and starch) have been shown to be the principal reserve

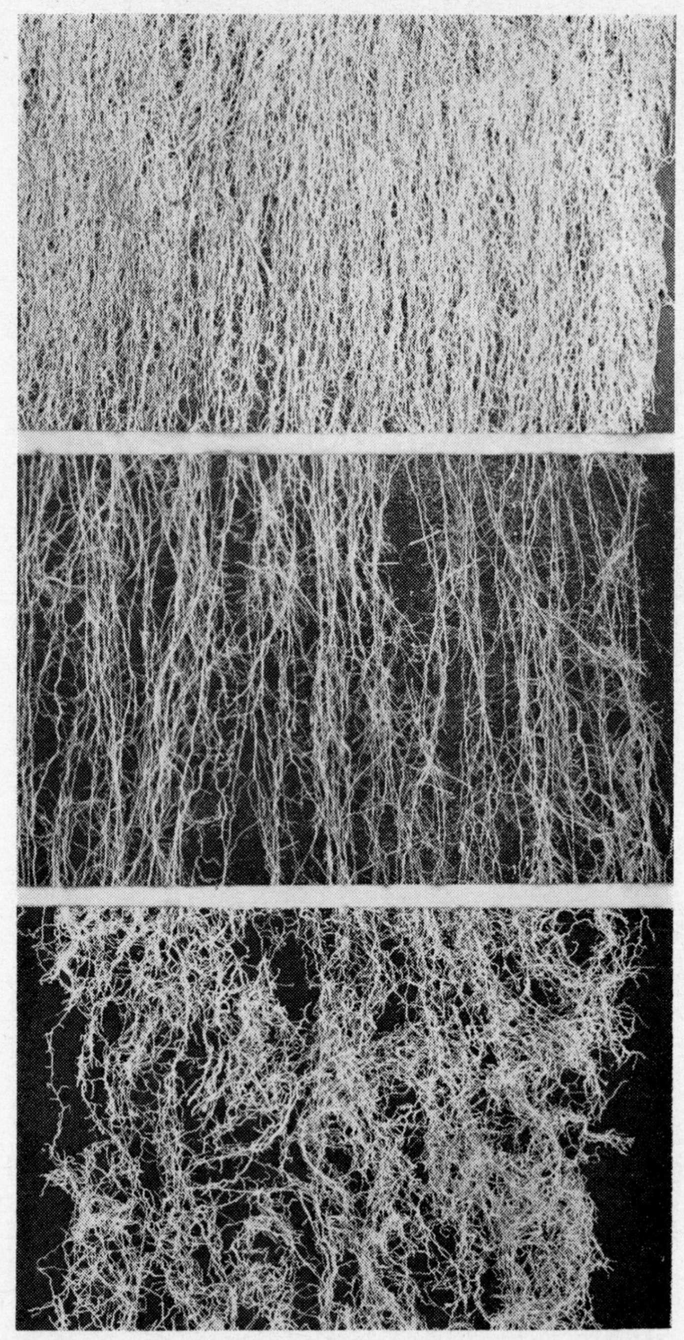

FIG. 51.—Portions of the root system of western wheatgrass taken from the A, B, and C horizons of a claypan soil near Lincoln. Explanation is in the text.

substances in grasses. These materials are elaborated by the leaves in excess after flowering, and are subsequently translocated to the roots and rhizomes, where they are stored to be drawn in the following spring for the production of new top growth. Nitrogen and mineral elements (though not being reserves in the true sense of the word but merely nutrients) are likewise translocated in autumn from the aerial parts to the underground system where they are stored over winter. This is one reason why most true prairie grasses have very little food value for livestock unless cut in late summer or early fall and stored as hay. It also accounts for the surprisingly rapid growth in spring of prairie cordgrass, big bluestem, and other rhizome-bearing grasses.

Prairie cordgrass has very coarse, much branched rhizomes which terminate in rigid, sharp-pointed buds. Some exceed one-fourth inch in diameter and may extend outward 12 to 18 inches before giving rise to erect shoots. Beneath a stand of this cordgrass, the soil to a depth of 1 to 6 or even 10 inches contains a network of coarse, woody, much branched underground stems. While mostly more or less horizontally placed, some branches are erect, and from these new rhizomes arise at various depths and connect several stem-bases. These yellowish to golden stems are both woody and flexible. They have a great tensile strength, resisting a pull of 50 pounds before breaking. Total length of the tangle of rhizomes was found to be 74 to 87 feet per square foot of soil surface. In Nebraska, thousands of acres of low, wet prairie land were occupied and protected from erosion by a wonderful stand of this grass.

Rhizomes furnish abundant room for storage near the food factories above them and are close to the water and supplies of nutrients around and beneath them. Fifty feet of rhizomes with an average diameter of less than one-fourth inch have a volume of nearly 12 cubic inches, in a large part of which food may be stored.

Switchgrass, which grows in somewhat drier soil than that of prairie cordgrass, develops a dense, entangled, and somewhat woody and much-branched mass of rhizomes as shown in Figure 52. These rhizomes are usually found at depths of 2 to 5 inches but sometimes they are 8 inches deep. They enable the plant, once established, to spread rapidly. Several rhizomes may originate from a single stem-base. All may give rise to new, erect plants. As in many grasses, new roots arise all along the path of the rhizome but they are thickest around the enlarged, vertical stem-bases. In old stands, rhizomes occurred at the rate of 50 feet per square foot of soil surface. Tensile strength of representative samples varied from 80 to 132 pounds.

FIG. 52 (upper).—Coarse rhizomes on the bottom of an 8-inch-thick mat of the underground framework of switchgrass on a flood plain. Under a square foot of soil there were 50 feet of rhizomes.

FIG. 53 (lower).—Surface view of a square foot of big bluestem after the soil was washed away and roots removed.

Big bluestem produces great numbers of strong, branched rhizomes which are compacted into dense mats about 2 inches thick at depths between 1 and 3 inches (Fig. 53). The tensile strength was 55 to 64 pounds for individual samples. Their length per square foot of sod averaged 55 feet. This indicates that an acre of big bluestem sod might contain 400 miles of rhizomes. Data on rhizomes and Figures 52 and 53 were originally published in a paper by the author in the *Journal of Range Management* of July, 1963.

About 22 species of grasses and grass-like plants, in addition to those described, are of considerable importance in lowland prairie. Some have been mentioned in wet land; others, such as bluegrass, side-oats grama, and Junegrass, will be described in the following chapter on upland prairie.

VI.

Grasses of Uplands

Upland prairie is quite different from that of lowland (Figs. 54, 55). The five dominant species, distributed in three plant communities, are all bunch grasses of medium height. In addition to the largest community of little bluestem, needlegrass, and prairie dropseed, each controls minor communities. Junegrass and side-oats grama, also dominant mid grasses, range widely throughout, while big bluestem, mostly in the bunch form, is intermixed to a moderate degree. Upland grasses cannot compete successfully with the tall grasses of lowlands because of the dense shade beneath them. There is, however, much intermingling of species, especially on lower slopes.

Little bluestem is by far the most important grass of upland prairie. On an average this grass alone composes about 60 percent of the upland vegetation and in places it furnishes 90 percent. In drier soils on the hilltops and upper slopes its bunch habit is pronounced, but elsewhere it forms an interrupted sod in which the mats and tufts are very dense. Its xeric nature is further revealed by its extending far westward into mixed prairie.

Needlegrass is typically an upland species. Steep, dry ridges and xeric slopes, especially where the soil is partly sand or gravel, are often more or less dominated by this grass. It is also more or less abundant on flat lands at the heads of draws and on broad washes of lower slopes where heavy rains may cause some erosion and deposit of soil. Its chief associates are Junegrass, to be described, and little bluestem. This cool-season perennial, which forms small bunches, is of boreal origin. It renews growth early in the season. Height of foliage ranges from 2 to 3 feet early in June; the abundant flower stalks are somewhat taller. Flowering follows rapidly and by June 10 the twisting of the 4- to 6-inch awns indicates the ripening of the seeds.

A fruiting field of needlegrass is a magnificent sight (Figs. 56, 57).

FIG. 54.—Loess hills at Allen, near Ponca, covered with tall-grass prairie. The dominant grasses are needlegrass, little bluestem, and big bluestem. Photo June 7, 1932.

FIG. 55.—Prairie on a hilltop near Nebraska City. This is a small part of a very large area. The most conspicuous plant is needlegrass, but other grasses were quite as common.

FIG. 56.—Example of native prairie with an abundance of forbs.

FIG. 57.—Needlegrass in fruit. The stems are not bent by the wind but by a heavy crop of seed. (From *North American Prairie*.)

The wand-like stems are often so thick, as they bend gracefully under the weight of the seed, and the foliage cover so dense that the general appearance, until one separates the crowns, is that of a sod former rather than a bunchgrass. Actually the basal cover is very small; only about 11 percent of the soil surface is occupied. After the fall of the seed, usually by the first week of July, the four-foot stems, now bleached white, become erect. They and the broad shiny glumes (flower parts) remain for several weeks and clearly demark the boundaries of the needlegrass community. The seed is buried in the soil by the torsion of the bent and twisted awn, which is held from turning by the bases of stems near the soil surface. Needlegrass is relished by livestock. Grazing animals feed upon it in early spring before other grasses have made much growth. It should not be mowed for hay until the seeds have fallen.

Prairie dropseed is a warm-season, perennial bunch grass. The foliage height is about 8 to 18 inches on upland but it is 6 inches taller on low ground. Growth of the narrow, yellow-green leaves begins rather early in spring from the bunches, which usually vary from 4 to 8 inches in size. Older clumps may be 12 or more inches in width. In good years the bunch is filled with densely crowded shoots but in drier ones their number is much less. The leaves, which are 18 to 40 inches in length and usually dry at the tips, curve gracefully outward and downward. Unlike most prairie grasses, they remain attached to the plant in winter and thus form a deep mulch. Flowering and seed production occur late in August or in September.

The root habits of representative grasses of uplands are shown in Figure 58. The average spacing of bunches, about 1 to 1.5 feet apart, should be noted. The largest roots have only about one-third the width of those of big bluestem. The soil beneath the crowns and on all sides of them is threaded with dense mats of roots to a depth of 4 to 5 feet. The average number of roots in the trench wall is shown except in the first two feet. Here they were even more abundant than in the drawings. Details of branching, of course, cannot be shown in so small a drawing. For example, the mature root system of little bluestem consists of a vast network of roots and finely branched rootlets, some of which are more than 30 inches in length and branched to the third order. Stem bases and surface roots firmly bind the soil in the mats or bunches. Almost all of the space between bunches is occupied by a dense network of roots.

One exhaustive study of the binding network of roots showed that a strip of prairie sod 4 inches deep, 8 inches wide, and 100

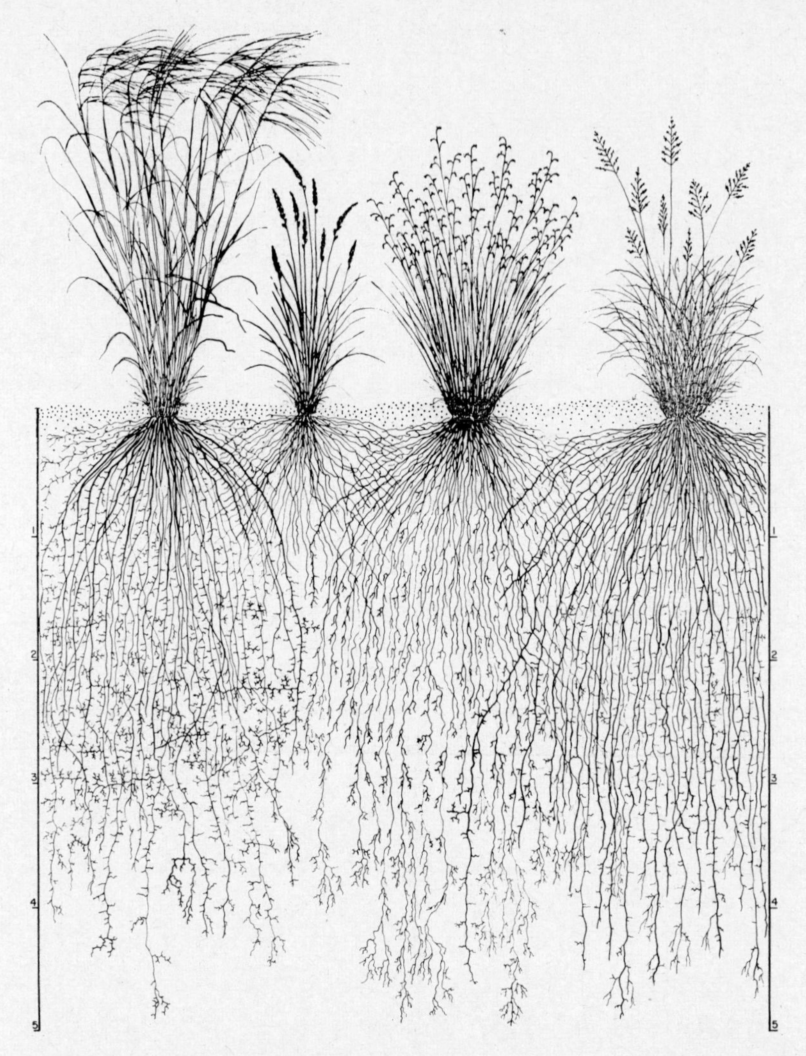

FIG. 58.—Characteristic development of top and roots of four bunch grasses as they occur in the several upland communities. From left to right they are needlegrass, Junegrass, little bluestem, and prairie dropseed. Note that the tops are only about half as high as the roots are deep.

inches long was bound together with a tangled network of roots having a total length of more than 20 miles. Moreover, the main roots have great tensile strength; individual roots of little bluestem are broken only under a pull of approximately 2.5 pounds. Strength of roots of needlegrass was only slightly less.

When one removes a bunch of grass from the sod and washes the soil from the roots, there remains a block of sod as shown in Figure 59. Here the surface roots have been cut off close to the bunch and then forced downward by the water in washing. The picture is thus misleading. In Figure 60 (center) blocks of sod were taken to include the soil 8 inches on all sides from the center of the bunch. The sod was then cut at a depth of exactly 4 inches. The soil was gently washed away after long soaking in water, thus exposing the roots that extended outward on all sides of the bunch. Several species of bunch grass were examined. All were similar to those in Figure 60. In washing, a considerable amount of other plant materials was separated from the roots and discarded. Many of the roots of prairie bunch grasses extend outward at least 2 feet in the upper 4 inches of soil, but most of them grow vertically downward.

A plot of little bluestem sod of about 6 square feet is shown in Figure 60 (lower). The larger bunches are little bluestem except one of Indian grass in the left background. Four small bunches are June-grass (in background) and needlegrass. Only the stem bases and roots that occurred in the upper 4 inches of soil are shown. Thus, 1 to 1.5 inches of stem bases occur just below the soil surface. Not included in the figure were the underground parts of various forbs, seedlings, and small clumps of sedges. These, when air-dried, weighed about a ton per acre. These data and Figure 60 were published originally in the *Journal of Range Management* in July, 1963.

Extensive studies have been made of numerous large samples of sod approximately 3 feet long and 1.5 feet wide but only 4 inches in depth. The air-dry weight of plant materials was 3.15 tons per acre in prairies of western Iowa, and 2.60 and 2.34 tons, respectively, in eastern Nebraska and central Kansas. The main cause of the great stability of upland prairie soil is the solidly anchored bunch grasses

FIG. 59.—Blocks of sod of prairie dropseed (left), needlegrass, and Indian grass showing stem-bases and roots in the surface 4 inches of soil.

FIG. 60.—Basal view (top) of bunches of prairie dropseed (left) and tall dropseed (right). Only the roots in the upper 4 inches of sod are shown. Plot of little bluestem prairie (lower) showing stem bases and roots in surface 4 inches of soil.

and other species and the interlocking of the tangle of tough roots between the bunches.

Hills carpeted with prairie are very slowly losing some soil by geological erosion. They are not eternal! They are far more vulnerable to erosion after the prairie sod is broken. The Blue River did not get its name because its waters were muddy, nor did the Missouri River—the Big Muddy—receive much of its sediments from the true prairie soil. Moreover, great dust storms do not originate in unbroken prairie. It is believed that during the historical period dust storms have come only from soil exposed by man in the course of settlement.

Besides little bluestem, needlegrass, and prairie dropseed, each of which form communities, numerous other grasses are common on uplands. Side-oats grama, a mid grass, is scattered widely throughout the prairie in all types of situations, but rarely occurs in great abundance (Figs. 61, 62). It nearly always grows as small, isolated, rather open tufts scattered among other species. It composes only 1

FIG. 61.—A bunch of side-oats grama in full bloom (left), and a similar bunch of prairie dropseed in September.

FIG. 62.—Many-flowered psoralea, a legume about 2 feet tall with widely spreading tops and a wealth of small, blue flowers.

Fig. 63.—Prairie near Pierce on September 15 after mowing and stacking the hay. The stubble and mulch give efficient protection to the soil. (From *North American Prairie.*)

to 3 percent of the vegetation generally but may occur as abundantly as 10 percent locally. Like blue grama and buffalo grass, which are found in certain situations in true prairie, it is far more abundant in mixed prairie westward. All of these grasses produce excellent forage.

Kentucky bluegrass is a rather constant component of true prairie both on uplands and lowlands where it makes up about 5 and 9 percent of the vegetation, respectively. It has spread very widely since the coming of the white settlers but is not a native American plant. Prairie fires were very destructive to bluegrass, but mowing aided it greatly, since with the removal of other vegetation it thrived both in early spring and late fall.

Two small, low-growing panic grasses rank high among secondary species. Scribner's and Wilcox' panic grasses fill interspaces between other species. Tall dropseed is a perennial, tufted, tall grass. It forms small clumps or occurs as scattered individuals in the sod matrix. Several sedges and a few rushes are also common. The foregoing species are examples of the 25 most important minor grasses of uplands (Fig. 63).

VII.

Forbs of Lowlands

Non-grasslike herbaceous plants or forbs are always components of prairie. Often they are more conspicuous, although nearly always of less importance, than the grasses. Species of forbs greatly outnumber those of grasses. In the western third of true prairie, where as many as 250 species of plants occurred in a single square mile, 142 forbs were found in at least 10 percent of the nearly 200 prairies examined. They occur naturally in two large groups of nearly equal size, those of greater ecological significance in lowland and those that are more characteristic of upland. Careful study of many prairies in Nebraska and parts of adjoining states during several growing seasons revealed that 67 species of forbs occurred in at least 10 percent of the low-lying grasslands. Seventy-five other species were found in all but 10 percent of upland prairies. This gives one an idea of their abundance and wide distribution, even if dozens of those that occur only occasionally or rarely are not included.

The prairies are not of recent origin. Their beginnings date back about 25 million years to Tertiary times. They resulted from the uplift of the Rocky Mountains and subsequent changes in climate, especially reduced precipitation. Present day prairie is preglacial in origin and has descended from the climatic prairie of the Tertiary period. Climax prairie is extremely resistant to invasion. Even small tracts of prairie, although surrounded by pastures or cultivated fields with their accompanying annual and long-lived weeds remain intact. Such stability denotes a high degree of equilibrium between native vegetation and its environment under the control of a grassland climate. All except a relatively few species of grasses and forbs in prairie are long-lived. No time is lost in spring in germination and establishment since nearly all of the plants are already wonderfully well established.

Forbs are an integral part of prairie vegetation; all add tone to the landscape, and the beauty of the prairie is in a large part a result of their presence. The list of native forbs eaten by livestock is a long one. Nearly all of the numerous legumes are valuable as forage, and they contribute considerable amounts to the total yields.

Forbs of lowland prairie in Nebraska are commonly taller and coarser than most of those in drier soil. Conspicuous among these are three species of rosinweeds (Figs. 64, 65). The foliage is coarse and abundant and the large composite, yellow flowerheads are very noticeable. Blooming occurs in both summer and fall. The cup plant is so called because of its large opposite leaves which are united at the base to form a cup which holds water after rains. The stem is square in cross section. This species is less abundant than the other rosinweeds. All attain a height of 5 to more than 8 feet.

The compassplant varies from 8 to 10 feet in height and under increased rainfall occurs on uplands as well as on low ground. It has very large, deep, and somewhat fleshy roots. Mature leaves are often 14 inches wide and nearly 2 feet long. They are not entire but deeply cleft. The vertical leaves, especially of young plants, have the orientation of a compass, the sides receiving the morning and afternoon sun. A single clump may cover a circular area 2 to 5 feet in diameter. From the root-crown and other thick underground

FIG. 64.—Three rosinweeds and a sunflower. Cup plant (left) and, in sequence, entire-leaved rosinweed, young specimen of compassplant, and saw-tooth sunflower, with base of its stem at right. (From *North American Prairie*.)

FIG. 65.—Jerusalem artichoke, another sunflower, with a quantity of enlarged underground stems. (From *North American Prairie.*)

parts, a single clump may give rise to 100 or more basal leaves which rise above the foliage of the grasses. The Indians and children of the early settlers gathered the gummy material which exuded from the flowers and upper part of the stem and used it for chewing gum.

A third species, entire-leaved rosinweed, often grew in clumps of 5 to 25 stems, and, like the preceding plants overtopped the grasses. This species and the compassplant ranked very high among other species of lowland.

There are three species of sunflowers which occur rather regularly in lowland prairie. They also belong to the very large family of composites—plants that have their flowers closely grouped in heads. The saw-tooth sunflower is so named because the edges of the abundant leaves are notched somewhat like the teeth of a saw. This plant is also abundant in ravines where it forms dense stands, and often migrates, during wet years, several yards into higher ground. Here the plants become more widely spaced on the rhizomes and much dwarfed. Renewed growth of this perennial from its thick crown begins early and proceeds very rapidly. A height of 6 to 9 feet may be attained in late summer, after which the yellow flowers become abundant.

Maximillian's sunflower is another coarse, tall species that grows in small clumps from rhizomes. It is far less abundant than the preceding. The Jerusalem artichoke is a lowland sunflower of minor importance but of much interest. It is characteristic of borderlands between prairie and shrubs and woodland and along ravines and streams. It is not so tall as the saw-tooth sunflower and its leaves are more like those of the cultivated species. The storage organs consist of tubers, sometimes an inch or more in diameter, which are often abundant (Fig. 65). These were eaten by the plains Indians raw, boiled, or roasted, but it is believed that they never cultivated them.

Water hemlock is a stout, erect, much branched perennial and poisonous plant that is found only in moist or wet soil (Figs. 66, 67). The base of the great, hollow, smooth stem is about an inch thick and is often marked with purple lines. Height varies from 3 to 6 feet. The petioles of the leaves are usually 1 to more than 2 feet

FIG. 66.—Water hemlock with long-petioled basal leaves and carrot-like flowers, and rhizomes and erect stems of American germander. (From *North American Prairie.*)

FIG. 67.—Showy goldenrod in late fall. It is also common on upland. (From *North American Prairie*.)

long and the outline of the compound leaf blade is usually more than a square foot in area. The small white flowers occur in umbels. The plant is supplied with several fleshy roots some of which may reach an inch in diameter.

There are numerous tall plants such as willow-leaved aster (3.5 to 7 feet high), tall goldenrod (3 to 6 feet), Sullivant's milkweed (3 to 4 feet), and blazing star (3 to 5.5 feet tall). Plants of intermediate height, such as mints, fringed loosestrife, pink phlox, and black-eyed Susan, are also common. Most species are defoliated in summer to a height of 6 to 24 inches because of the dense shade produced by the tall grasses.

In spring and early summer one finds numerous low-growing plants. The scarlet strawberry is widely distributed. It spreads mostly by long runners (stolons) although it also has short, thick, shallow rhizomes. The pretty white flowers are most abundant in May, and large clusters of bright red, delicious fruits ripen during June.

Canada anemone is widely distributed and very conspicuous in the wetter portions of big bluestem prairie, where the grasses are not too dense, and in depressions and sloughs. Here, resulting from its propagation by rhizomes, it often forms dense, more or less circular patches a few to many feet in width. The abundant, large, white flowers unfold during the latter half of May and there is a splendid display of blossoms for several weeks.

Scattered throughout the prairie are various violets, spiderworts, purple and yellow sheep-sorrels, and, less frequently, the bright flowers of yellow stargrass. American vetch, with its purple pea-like flowers, occurs in patches, and prairie cat's-foot is often present in great abundance. In a study extending through several growing seasons, about 200 species of forbs most characteristic of lowlands have been observed. Each species has its place among the dominant grasses.

The group of tall species, which carry most of their leaves above the grasses, composes 50 percent of the lowland forbs. Forbs of low stature, which rarely exceed a height of 1 to 1.5 feet, furnish only 19 percent. A third group, which makes its best growth in spring and early summer, is composed of plants of intermediate heights of 2 to 2.5 feet, which is approximately the midsummer level of the grasses. This group makes up the remaining 31 percent of the total. Examples of this group follow.

The fringed loosestrife has a very wide distribution and was found in every lowland prairie examined. It grows in dense stands from underground stems. The leaves are widely spaced on the unbranched stems, and the plant is very tolerant of shade. The pretty yellow flowers are held quite above the foliage.

The golden meadow parsley forms societies in about half of the lowland prairies. It is an erect, smooth, branched perennial. The plants sometimes form rather dense groups, since they possess thick, short, underground stems. Although there are only one to eight stems per plant, the very large, much-compounded and abundant leaves give the plant a bushy appearance. The flower buds appear early and the yellow-flowered, carrot-like umbels are very conspicuous during May and June.

Prairie phlox is a perennial. The large clusters of beautiful pink-purple flowers are a pleasing feature of the landscape in May and June. The slender, erect, mature stems for a long time stand well above the grasses. Where rainfall is plentiful this phlox also grows on uplands.

Other species in this group are stiff marsh bedstraw, scouring rushes, and several mints. The preceding species give only a brief glimpse into nature's wonderful flower garden—lowland prairie (Figs. 68, 69).

Fig. 68.—Nest of a meadow lark somewhat opened to show interior. Most of the debris on the soil surface was left by the mower. It is early spring and only a few green shoots have appeared.

Fig. 69.—Rattlesnake of lowlands in eastern Nebraska.

VIII.

Forbs of Uplands and Viewpoints on Prairie

A five-year study of the general distribution of upland forbs revealed that 75 species occurred regularly in all but 10 percent of the prairies examined. Total upland forb population, including all species encountered, was approximately 200. Only a few can be shown here, but many are described and illustrated in *North American Prairie*. In general, upland forbs are not so tall as those of lowland, nor is their growth in height so rapid. This seems to result from the less dense shade which they encounter and perhaps the drier habitat.

The most abundant species and the one with the most consistent distribution is lead plant (Fig. 70). It is considered here as a forb although it is a half-shrub kept at low level because of annual mowing. In a very large number of prairies it formed the most important and most extensive societies. In mowed prairie the plant develops 3 to 5 or more stems which by late summer become quite woody. Flowering begins late in June when the plant is about 2 feet tall and thickly covered with the leaden-colored leaves. A maximum height of 3 feet may be attained before autumn. The dark purple or indigo flowers are conspicuous for several weeks. The flower has only a single petal which is wrapped around the stamens and style. Competition of lead plant with grasses for water is greatly reduced because of its great depth. The extensive, woody root system is 6.5 to 16.5 feet deep. Nodules occur throughout the entire root-length of this legume. Many of the tough, woody roots spread more or less parallel to the soil surface at shallow depths. In breaking the sod these broke with a characteristic snap thus giving rise to the colloquial name of prairie shoestring.

The prairie flora includes several half-shrubs such as prairie rose, dogwood, and redroot or new Jersey tea. Redroot, with its inch-thick taproot, gave the sod-breaker even more trouble.

FIG. 70.—An upland prairie near Lincoln with an abundance of lead plant or prairie shoestring. Self-recording instruments measuring soil and air temperature and humidity are in the shelter boxes.

Missouri goldenrod is the most widely distributed and most important of the numerous goldenrods of prairie. It is a coarse perennial that spreads widely by means of stout rhizomes; mature plants among the grasses attain a height of 1 to 3 feet. The woody base develops numerous rhizomes often one-fourth inch in diameter and 2 to 12 or more inches long. Many roots arise from the shallowly placed rhizomes and clusters of them from the base of the stems. Some extend downard 5 to 8 feet. Nearly all of the many species of goldenrods have rhizomes. The stems are usually not densely clustered. Those of large plants are somewhat woody. Early in May the new foliage stands out prominently above the grasses. Where plants are abundant the density of the grass is considerably reduced. Blossoming begins late in July with the beginning of the autumnal aspect. Then the prairies are aglow with the yellows of various goldenrods and sunflowers and, a little later, the purple of the blazing stars.

Prairie mugwort is conspicuous because of the grouping of plants resulting from propagation by rhizomes and its light gray appearance against the background of green (Fig. 71). Nearly all of the rhizomes occur in the upper 4 inches of soil and are joined in a complex network. Individual rhizomes are often 2 feet long. Mature

FIG. 71.—Prairie mugwort and various upland grasses in midsummer.

plants are about 2.5 feet tall with tops much branched. The flowers are small and not very conspicuous. This sage is rather local in occurrence but where it is found the plants are often densely aggregated and the grasses are thinned.

Plants of many-flowered aster, which often form bushes about 3 feet high, are also connected by tough, somewhat woody rhizomes. These are only about 1 to 8 inches long. The stems are densely matted together with fine fibrous roots, some of which are 7 to 8 feet deep. However, numerous fine and shorter roots absorb mostly from the surface soil. This species is one of the three most abundant and widely distributed of upland forbs. It is a perennial and makes a vigorous new growth in April, and the stems stand well above the grasses in early May. It is especially conspicuous from the time when its small white or purplish flowers begin to appear, late in August or early September, until the seed is ripe. Often the flowers are so densely clustered that they almost obscure the rest of the plant.

Stiff sunflower is one of the most widely distributed and most abundant and characteristic plants of upland prairie (Figs. 72, 73). It also occurs on about half of the lowlands. New shoots develop rapidly mostly from the widely spread and abundant rhizomes. Growth begins early in spring and the plant soon exceeds little bluestem in height. In dry soil a height of only 12 to 18 inches may be attained and the flower heads may be restricted to one per plant. But if water is more plentiful the plants may attain a height of 3 feet and the size of both leaves and flower heads may double. Even then the heads are only 2 to 2.5 inches in diameter; the lower leaves become dry; and a few pairs of upper leaves constitute the entire foliage. Blossoming begins early in July and continues until mid-September. Although the plants may grow thickly they do not produce much shade. Three flowers per stem is common.

Prairie cat's-foot is a low-growing mat former. Some of the leaves remain alive all winter and new ones appear in mid-March. Early activities are directed toward the production of flower stalks. Most of the new leaves and the abundant stolons are developed after the period of flowering. This plant is dioecious; usually the small patches are composed of either staminate or pistillate plants. The flower heads are borne on erect almost leafless stems 2 to 6 inches high. Dozens of these may occur in a single square foot. Only the stems bearing seeds elongate and thus place the tiny, bristly fruits in a favorable position for wind distribution. The whole appearance of all parts of the plant is dull gray. It occupies the interspaces between other plants.

FIG. 72.—Stiff sunflower in midsummer (left), and tops and enlarged roots of prairie turnip. (From *North American Prairie*.)

Fig. 73.—Ground plum with an abundance of fruit. (From *North American Prairie.*)

The list of prairie forbs is a long one. It includes the following: prairie clovers (white and purple), ground plum, Indian turnip, purple blazing stars, black-eyed Susan, falso indigo, Pitcher's sage, prairie rose, blue-eyed grass, yellow flax, sages, and dozens of others (Figs. 74, 75). Many are illustrated and described in *North American Prairie.*

Roots Systems of Forbs

The roots of various plants have been studied by the writer over a long period of years. That forbs penetrate through the network of fibrous roots of grasses and grass-like plants is shown in Figure 76. In studying underground plant parts, it is first necessary to excavate a deep trench in the walls of which one may examine the root system by means of a hand pick and an ice pick. It is the usual procedure to draw the roots to scale in one plane as they are exposed. Such a chart of blazing star is shown in Figure 77. Sometimes it is possible to excavate an entire root system which may then be photographed. Since each drawing or photograph usually requires an entire page when reproduced, a new method of showing root systems has been devised. When one has worked with root systems for

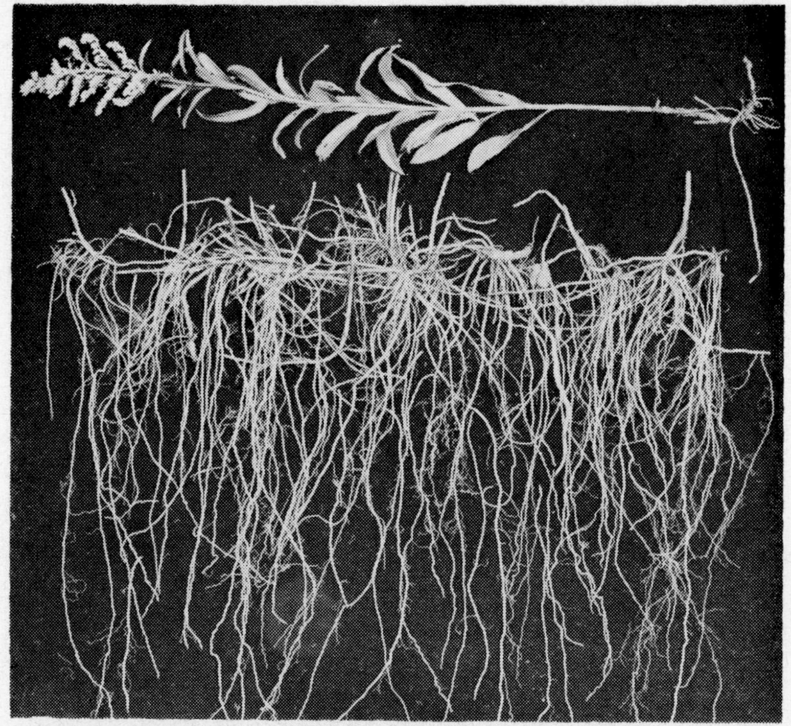

FIG. 74.—Missouri goldenrod with rhizomes and surface roots. It often forms large bunches 2.5 feet tall with a root-depth of 7 to 10 feet.

FIG. 75.—Rhizome system of prairie mugwort with tops and roots removed.

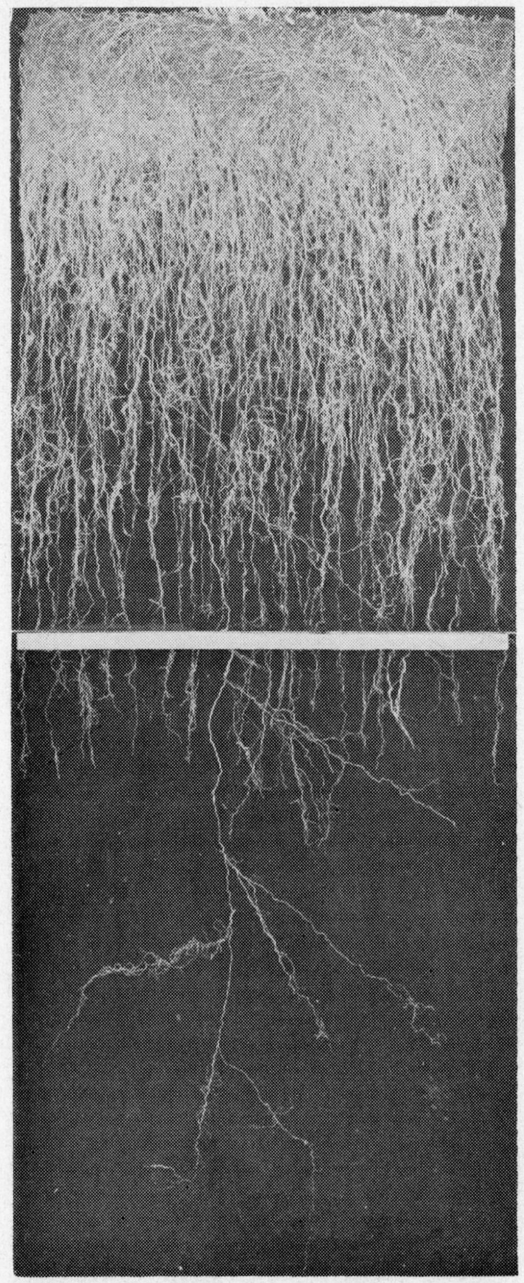

Fig. 76.—Roots of little bluestem from a monolith of soil 2.5 feet wide, 3 inches thick, and 6 feet deep. It was taken in two 3-foot sections. A branch root of a prairie legume (many-flowered psoralea) penetrated through the sod and to a depth of nearly 6 feet.

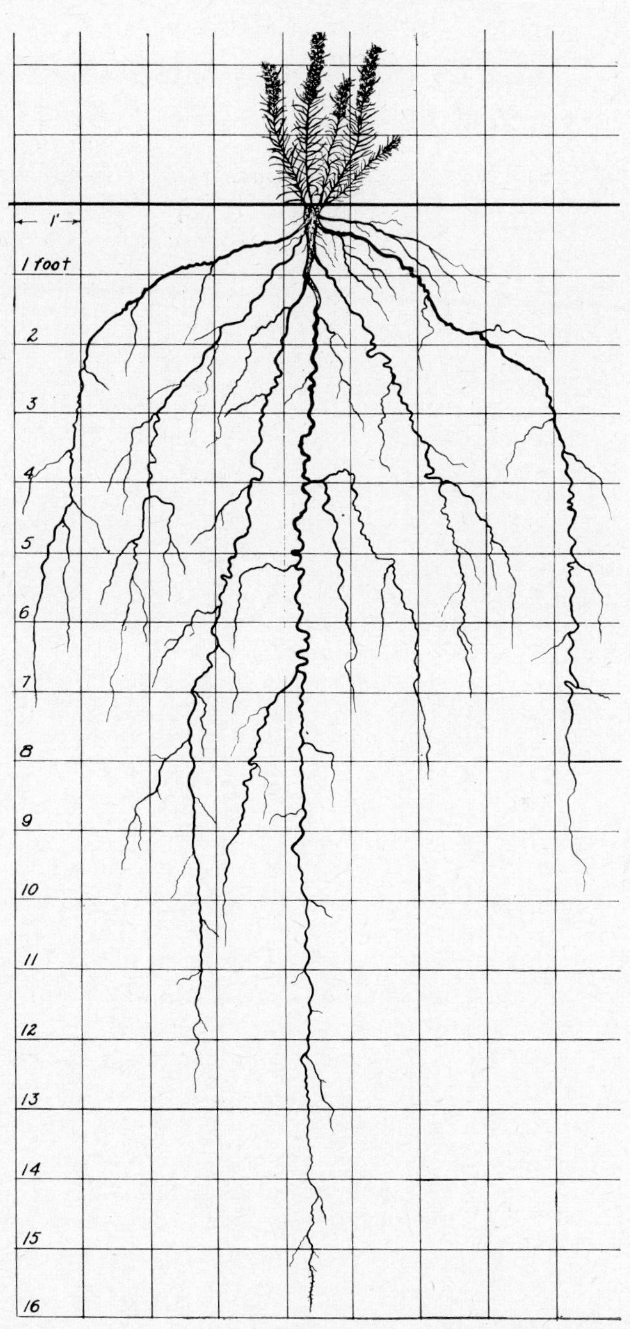

Fig. 77.—Root system of a blazing star with a lateral spread of 4.5 feet and about 16 feet deep. Roots of some prairie forbs extend to greater depths but few if any have a greater lateral spread. (From *North American Prairie*.)

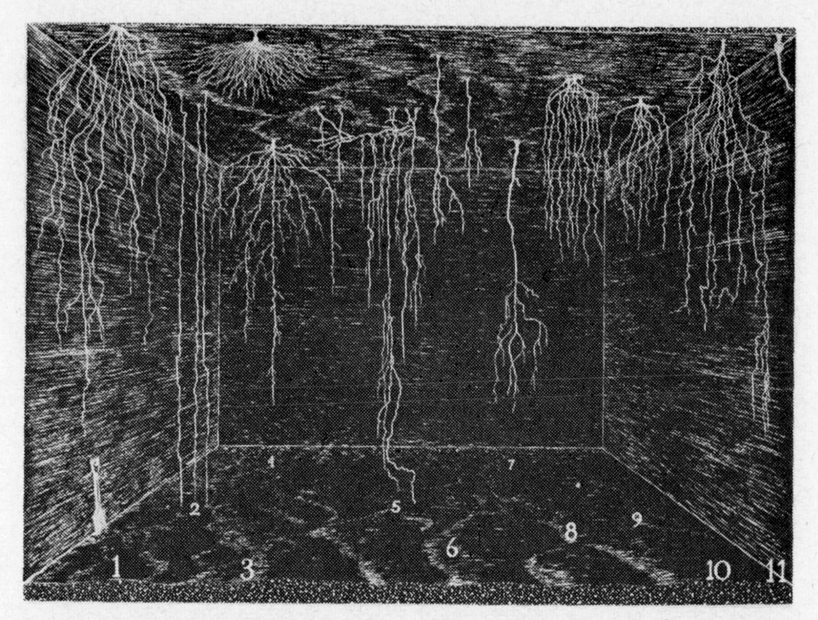

FIG. 78.—Excavation 20 feet deep showing root habits of several prairie forbs. 1, lead plant; 2, rush-like lygodesmia; 3, scaly blazing star; 4, dotted button snakeroot; 5, prairie rose; 6, pale purple coneflower; 7, compassplant; 8, many-flowered aster; 9, ground plum; 10, licorice; 11, prairie turnip, turning aside into wall. (From *Botanical Gazette*.)

several years, he can visualize, when standing in a deep trench, what will be seen when the soil has been removed. All of this mental image can be shown in a single picture. Figure 78 is a sketch of such an image. For clarity, the network of grass roots has been omitted and only an inch or two of the topsoil left in place. The figure shows the roots of ten different forbs in an excavation 28 feet wide, of similar length, and 20 feet deep. The root systems are of natural width and length and are representative of numerous prairie forbs. When the picture is inverted, one gets a better idea of the considerable space between the forbs which, of course, is occupied by grasses. The abundant, finer branches of roots could not be shown.

A dozen or more forbs in an area this size is common in prairie, although all of these species may not occur in such a group. The root drawings are from publications of the Carnegie Institution of Washington. Others will be shown for sand hills and mixed prairie of the Great Plains.

The beauty of the prairie is so impressive, as one wave of blossoming plants is followed by another from early spring to late autumn, that the seasonal aspects have been described separately in *North American Prairie* (Figs. 79, 80).

VIEWPOINTS ON PRAIRIE

It is of interest to consider the viewpoints of the prairie as re-

FIG. 79 (upper).—Wild strawberry (left) and blue-eyed grass in spring.
FIG. 80 (lower).—Black-eyed susan in summer (left) and Missouri goldenrod in fall. (Photos from *North American Prairie*.)

corded by a layman and by several scientists. At a very early date (1839) a description of the Iowa prairies was given as follows by Judge J. Hall:

> The scenery of the prairie is striking, and never fails to cause an exclamation of surprise. The extent of the prospect is exhilarating. The verdure and the flowers are beautiful, the absence of shade, and consequent appearance of light, produce a gaiety which animates the beholder. The whole of the surface of these beautiful plains is clad, throughout the season of verdure, with every imaginable variety of color, from grave to gay. It is impossible to conceive a more infinite diversity, or a richer profusion of hues or to detect the predominating tint, except the green, which forms the beautiful background, and relieves the exquisite brilliancy of all the others.*

Dr. B. Shimek, who knew the Iowa prairies from boyhood and served as a professor at the University of Iowa, described them as follows in 1911:

> The prairies presented varying aspects. The early settler avoided them at first in part for the reason that he thought them not fertile because treeless, and in part because they did not furnish the much needed building materials, fuel and water; but as his experience increased, there were added to these reasons the menace of the prairie fires and the terror of winter storms. The old-time blinding blizzard of the prairies (with settlement) lost its horror, and though it may still cause personal discomfort, it no longer menaces the safety of its hapless victims. But even in winter the prairie was often attractive, for the storms subsided, and by day the sunlit sea of snow sparkled with countless ice-crystals which covered its surface, and by night it rested in impressive silence under the star-spangled sky. But the horrors and the charms of winter finally passed, and with opening spring the ponds and lakes were soon gilded with the water crowfoot, and the hills and higher prairies were dotted with the early pasque flower, the prairie violet and a variety of rapidly succeeding spring flowers. Soon the grasses covered the surface with a great carpet of green painted with puccoons, prairie phlox and other flowers of late spring. But the real rich beauty of the prairie was developed only after mid-summer when myriads of flowers of most varied hues were everywhere massed into one great painting, limited only by the frame of the horizon, uniform in splendid beauty, but endlessly varied in delicate detail.**

*J. Plumbe, Jr., "Sketches of Iowa and Wisconsin," *Annals of Iowa,* 14 (1839—reprint 1925) 483–531, 595–619.

**B. Shimek, "The Prairies," *Iowa University Bulletin from the Laboratory of Natural History,* 6 (1911), 169–240.

In 1944, Dr. B. E. Livingston, from the east and a professor of botany, explained his experience:

> A person who has just migrated from a forest region to the open country of a typical prairie is likely to find the latter unattractive. The wide and uniform horizon may seem, curiously enough, to shut one in, as under a tightly-fitting bowl of sky. The immediate foreground seems depressingly monotonous, for it offers but very little of variant detail to the inexperienced eye. With closer acquaintance, however, endurance gives place to affection. The sky-bowl, no longer confining, invites the imagination outward in free sweep. One learns to see and recognize a constantly enlarging myriad foreground of details. Plants, rodents, insects, all are forever in motion, and the picture changes fascinatingly from day to day and with the march of the seasons. There is indeed no limit to the interesting and inspiring happenings of a stretch of natural prairie, nor to the questions that arise concerning the how and when and why of prairie nature.*

After a half century of study the writer maintains his views as stated in 1934 and summarized in his book, *North American Prairie* (1954):

> The prairie covers a vast area. It appears almost monotonous in the general uniformity of its plant cover. The absence of trees, the paucity of shrubs and half-shrubs, the dominance of grasses, and a characteristic flora constitute its main features. Prairie is composed of many different species of native American plants. It appears as an inextricable mass of endlessly variable vegetation. One glories in its beauty, its diversity, and the ever changing patterns of floral arrangements. But he is awed by its immensity, its complexity, and the seeming impossibility of understanding and describing it. But after certain principles and facts become clear, one comes not only to know and understand the grasslands but also to delight in them and to love them.
>
> The prairie is an intricately constructed community. The climax vegetation is the outcome of thousands of years of sorting and modification of species and adaptations to soil and climate. Prairie is much more than land covered with grass. It is a slowly evolved, highly complex, organic entity, centuries old. Once destroyed, it can never be replaced by man.

*In preface to J. E. Weaver, "North American Prairie," *The American Scholar* 13 (3) (1943), 329–339.

IX.

Soils

Many years of investigation have permitted thorough examination of grassland above, in, and below the soil. The wonderful network of roots in the prairie sod has clothed and protected the land for untold centuries. We have seen the intimate structure of this network of roots and rhizomes which characterized the prairie over thousands of square miles. The root network is a permanent feature of prairie. Roots of dominant grasses and of many forbs are long lived; death and decay proceed slowly as does their replacement by new ones.

Soil may be defined as the unconsolidated outer layer of the earth's crust which, through the processes of weathering and the incorporation of organic matter, becomes adapted to plants. In prairie and plains the soil usually contains and acts upon a much more extensive portion of the plant body than does the atmosphere. Perhaps the greatest influence of climate on soils is exerted indirectly through its partial determination of the kind of vegetation under which the soil evolves. The true soil or solum is usually made of the parent materials which it covers.

The topsoil or A horizon in Brunizem soils of eastern Nebraska prairie is usually about 18 inches thick. Because of a relatively high precipitation, 25 inches or more under which it developed, the mellow surface soil has a granular structure. Repeated freezing and thawing, alternate wetting and drying, together with a greater humus content and the favorable effects of root activities, have combined to produce the excellent granular structure.

The subsoil or B horizon usually extends to a depth of about 30 inches. It has a higher clay content and a prismatic structure (Fig. 81). Root penetration is much more difficult and branching is less pronounced. Below the subsoil is the C horizon of parent materials from which the soil developed. Here roots penetrate much more easily. Since clay is less abundant, the material has a moderately

FIG. 81.—A Brunizem soil profile in silty clay loam near Lincoln. Depth of the A horizon is 13 inches; the prismatic B horizon occurs between 13 and 42 inches; and only the top of the C horizon or parent material is shown. Photo John Elder, Soil Conservation Service.

mellow consistency. Under Nebraska's prairie the parent materials are primarily deposits of wind-blown loess and glacial drift, but in a few places they are limestone or sandstone or beds of sand, clay, or volcanic ash. In a few places the loess is 50 or more feet in depth;

in others it has been removed from underlying glacial deposits or become intermixed with them.

The soil over most of the area is slightly acid since lime has been leached from the surface and subsoil. The deeply rooted grasses absorb much lime from the C horizon and upon the death of the tops each year some lime is returned to the soil surface. Soil scientists have ascertained that the dark humified organic matter ranges from 30 to 40 tons per acre in the top six inches of soil. It furnishes a home and food for hordes of microscopic plants and animals that prepare the soil for use by higher plants and a reservoir to supply plants with needed moisture. Here also nitrogen is stored in the organic material until it is made available by microscopic organisms for use of higher plants. The favorable climate and high fertility make the Brunizem soils very productive for native grasses and cultivated grass crops, such as corn and wheat.

Brunizem soils occupy only a part of Nebraska, about two counties wide west of the Missouri River, but only a narrow extension beyond 42° N latitude, i.e. Thurston County (Fig. 6). They give way on the west to a very wide and continuous belt of Chernozem soils lying northeast, east, and southeast of the great central Sandhill area. Chernozem soils develop under a continental climate of less than 25 inches rainfall and usually an excess of evaporation, as measured from a free water surface, over precipitation. Their eastward part in Nebraska is clothed with true prairie, while under lighter rainfall westward mixed prairie prevails.

Chernozem soils are thus found in the more humid part of the drier region. The luxuriant growth of grasses has produced black soil very high in content of organic matter and very fertile for cultivated cereal crops, especially wheat, and sorghums.

The introduction of living material is largely responsible for constructional processes of soil development, since residues of plants and animals return to the soil more than green plants take away. The soil now contains stored up energy and becomes the abode of bacteria and other microorganisms. Raw plant and animal wastes are converted into plant nutrients and dark colored matter of high carbon content by the activities of the microorganisms. Total organic matter usually constitutes 1 to 5 percent of the dry weight of true prairie soil.

Soil scientists have summarized the major soil-forming processes in Brunizems as the accumulation of organic matter in the surface layer; leaching of bases (lime, etc.) and development of acidity in the A and B horizons; the formation of a high cation-exchange capacity clay; and the accumulation of this clay in the B horizon.

Soil forming processes leading to the development of Chernozems are mainly the addition of organic matter to the surface horizon; the formation of clay and development of structure in the subsoil; and the movement of bases by leaching to a zone of accumulation in the lower part of the solum.

The root network is a prominent feature of prairie soils. Samples taken in spring, summer or fall are the same. While most of the changes resulting from plant growth have occurred in the soil, in the parent materials many changes also result from the presence of roots. This condition prevails even where the solum is several feet thick. Aside from the roots of grasses which often penetrate far beyond the solum and to depths of 8 to 10 feet, many roots of legumes with abundant nodules and roots of numerous other prairie forbs occur, some to depths of 15 to 20 feet. Plants take up lime, potassium, phosphorus and other nutrient materials from great depths and after death leave some of these in or on the solum. Earthworm burrows occur in parent materials of prairie soils, sometimes to depths of 13 feet. When roots decay, pores and channels are formed which accelerate the entry of air and water and thus increase chemical weathering. Hordes of organisms find entry along old root channels and cause decay. Upon their death they add to the considerable amount of decayed roots. In these ways soil and plant growth are closely related even in the C horizon.

The physical factors of the soil environment of the prairie have been measured during the growing season for a period of twelve years beginning in 1915. Determinations were made on both upland and lowland soils in extensive areas of unbroken prairie at Lincoln. The former was dominated by little bluestem and the latter by big bluestem. The soil is a deep, fertile, fine-textured silt loam of high water-holding capacity.

The mean annual precipitation is 28 inches, of which nearly 80 percent falls during the growing season. Periods of drought are liable to occur at any time and especially after midsummer. Water-content in the surface six inches of upland soil varied widely and rapidly, often 10 percent or more during a single week. Water available to plants was reduced to less than 5 percent one to four times during 11 of the 12 years. Only twice during this period was the water-content reduced to the nonavailable point.

Available water-content in the 6 to 12 inch soil layer exceeded 5 percent three-fourths of the time, but fell to 2 or 3 percent at 17 different intervals. At no time was the water available for plant growth entirely exhausted. In the second, third, and fourth foot the water-content was less variable. In general, there was a gradual decrease in the supply with the advance of the summer. This was frequently temporarily interrupted, especially in the second foot, by heavy rains. The available supply usually ranged between 5 and 15 percent. The maximum was 21 percent and a few times the minimum fell to 2 to 3 percent.

On the lowland available water-content was 3 to 10 percent greater in the surface foot and often 5 to 11 percent in excess of that of the upland in the deeper soil. A close positive correlation was found between precipitation and water-content, especially in the surface six inches.

Average day air temperatures sometimes reached 90° F. but were more usually between 75° and 85°. They were usually 10° or more higher than those of the night. Soil temperatures showed a daily variation of 15° to 18° F. at 3 inches depth but only 1° to 3° at 12 inches depth. The temperature decreased regularly with depth. Temperatures of air and soil during the long growing season are well within the ranges critical for plants of the prairie and are probably of secondary importance.

The average day humidity in the prairie varied between 50 and 80 percent during years of greater rainfall but fell frequently to 40 to 50 percent during drier years. The average night humidity was frequently about 20 percent higher. Both showed weekly ranges of 8 to 20 percent. No consistent differences in humidity were found throughout the three summer months. The humidity on low prairie was usually 5 to 10 percent greater than on the upland. Low humidity nearly always occurred during periods of low water-content of soil and usually also during periods of high temperatures.

Wind movement was fairly constant just above the grass and often high. It is an important factor in promoting water loss. Evaporation varied greatly from year to year. High evaporation was correlated with low humidity and both of these with low water-content of soil. Water-content of soil and humidity are the master factors of the environment of the prairie. The climax vegetation is remarkably well adapted to these water relations.

X.

Transition to the Great Plains and Loess Hills

As one travels westward across the Nebraska Plain and then enters the Loess Plain Region (Fig. 7) he is impressed by several changes. These changes result from the gradually increasing unfavorable environment as the vegetation of true prairie gives way to that of mixed prairie. The average decrease in rainfall from east to west across the state has been computed as one inch for each 25 miles. Conversely, rate of evaporation increases. Extensive studies through Thayer, York, and Pierce Counties revealed that the vegetation was that of true prairie. But in Phelps and Harlan Counties (Holdrege area) typical mixed prairie prevailed. It is in the intervening area, about 60 miles in width, that the transition occurs.

It was found in the studies that as one proceeded westward towards this area of transition, the grasses become shorter and more mesic species did not extend so far up the slopes. Big bluestem and switchgrass became much less widely distributed and showed no marked tendency toward replacing little bluestem, as occurred eastward. The sod formed by big bluestem was much more open and other vegetation was not so completely excluded by the shade. Yields of hay became smaller as the cover became more open and the grasses more dwarfed. Little bluestem remained dominant over much of the terrain where one would ordinarily expect to find big bluestem.

In the area of transition blue grama and buffalo grass gradually appeared between the bunches of little bluestem and the short grasses became intermingled as an understory to little bluestem. Other changes occurred gradually but widely. Density of short grasses between the bunches of little bluestem and certain other true prairie grasses greatly increased and gradually, under continued grazing, the short-grass pastures were quite in contrast to the bluegrass pastures eastward. In these dry prairies, whether level or rolling, switchgrass, nodding wild-rye, and prairie cordgrass found suitable habitats quite limited.

At stations bordering the transition zone on the west, the dominance of grasses had rather completely changed. Blue grama and buffalo grass were the most important and occurred in continuous and sometimes nearly complete stands. Elsewhere they were found as an understory to side-oats grama, western wheatgrass, needle-and-thread, June grass, sand dropseed and numerous other grasses of mixed prairie, to be described. Big bluestem, Indian grass, and other tall grasses were often found in deep, moist ravines. Thus, the dominant mid grasses were of a more xeric type, and the prominent understory of short grasses, not found in true prairie, covered much of the soil without the presence of a good overstory of mid grasses.

The increased xerophytism from east to west was revealed by the forbs by a decrease in the number of species, smaller stature, and the presence of many Great Plains species. In true prairie of eastern Nebraska an ordinary native area of 40 to 60 acres in extent possessed from 65 to 90 species of forbs. But only 25 to 40 species were found in areas of similar size in the western part of the transition.

The western prairies were not especially conspicuous, as were

FIG. 82.—View of the loess hills near Kearney. Cattle raising is a chief industry. (From *Grasslands of the Great Plains.*)

those of eastern Nebraska, because of the forb population. This was partly due to the dwarfness of the plants. Often they attained no more than half the height of similar species in eastern counties of the state. Many of the taller and more conspicuous forbs found eastward did not occur or occurred much less frequently. Those that did occur were not only of less abundance but were also greatly reduced in size. A long list might be given. Conversely, many western species, which were rarely seen eastward, were present in the area of transition. If the middle of the zone of transition were to be indicated by a single line, one (98° 30′ west longitude) extending from O'Neill in eastern Holt County southward through Greeley, Hall (Grand Island) and Webster Counties would be representative.

North of the Platte Valley Lowland (Fig. 7) the transition from true prairie to mixed prairie is somewhat different since it is over high hills and intervening valleys (Fig. 82). Here species of bluestem prairie have migrated along the valleys to the sand hills. Short grasses extended eastward along the dry hilltops, while a third community of intermixed species from eastern prairie and Great Plains clothed the midslopes.

Origin of Loess Hills

In central Nebraska southeast of the sand hills is a vast area, several thousand square miles in extent, of bluffs and valleys and uneroded uplands known as the Loess Hills and Plains (Fig. 83). About half of this region is still clothed with native grasses. It has been pointed out by geologists that the Sand-hill Region has released much Tertiary and Pleistocene loess-forming material which was blown prevailingly eastward and southeastward contributing directly or indirectly to Loveland and Peorian deposits of loess in a broad area east of the sand hills. They also find that most loess deposition in Nebraska correlated with interglacial time and that wind, correlated with other factors, was the major force in the genesis of loess. The aggregate thickness of the loess mantle approaches 150 feet or more in places.

Headwater erosion into the areas of thick loess mantle rapidly developed the characteristic companion topography of the region. As soon as the sod of the upland plain is removed by undercutting, the relatively coarse materials of loess are subject to rapid erosion by water because of the underlying silt texture and lack of cementation. Nearly vertical canyon walls result. Moreover, weathering of the canyon walls tends to vertical or nearly vertical cleavage in the loessial materials forming lines of weakness along which the loess tends to slide downward. This results in catsteps which are so typical of the side slopes of the upland of this region (Fig. 84), and which affect distribution of vegetation.

FIG. 83.—View in the northwestern portion of the dissected loess plains showing the bluffs and rolling land characteristic of the area. The loess soil is protected by a thick carpet of blue grama, a cover of wheatgrass, or by other vegetation. Photo southeast of Broken Bow.

FIG. 84.—View of catsteps caused by slipping of the soil.

The light yellow to brownish yellow loess is composed largely of particles of silt. It is uniform in texture, mellow, and very deep. The top portion of mature soils is dark brown and usually ranges from 8 to 12 inches in thickness. Water penetrates easily, lime content is high, and the carbonates are usually leached to a depth of 3 to 6 feet.

Immature soils are present over a large part of the uplands. Erosion prevents the accumulation of much organic matter and the topsoil is thin and light in color. It is usually 4 to 8 inches thick but may be much less. There is no zone of lime accumulation and the surface soil gives way directly to parent materials. The erosiveness of these fine textured soils is great. There is some constant erosion by wind and water from the steep bluffs and hills, but when the cover of grass is broken by grazing or trampling or otherwise, erosion immediately becomes serious (Fig. 85).

The greater part of the upland plain, because of stream and wind erosion, is now a hilly region. The average elevation is about 2,200 feet above sea level. It is traversed by numerous low-lying strips of flat alluvial land along rivers and larger streams. Not only

FIG. 85.—Deeply eroded cow paths in loess. Dark places in path on left are 40 inches deep. The second path is 11 inches deep and is being abandoned for a new one on the right.

is the greater part of the region well drained, but over large areas runoff is excessive and erosion is severe.

The mean annual precipitation is 23 inches at Broken Bow in the northwest and 24 inches at Kearney in the southeastern edge of the loess hills. Its distribution is of the Great Plains type; nearly 80 percent occurs between April first and September 30. Often the rain is very heavy over a short period of time. In May and June periods of drought are uncommon, in July rainfall distribution is less favorable; and during August and September long periods of drought sometimes cause reduced yields even of the crop of native grasses. Wind movement is fairly constant and often high. It is an important factor in promoting water loss. Mixed prairie is typically a land of sunshine, and the proportion of clear days is high. The climate is well suited to the production of hay and pastures, grain crops, and the raising of livestock.

As pointed out in *Grasslands of the Great Plains,* the mixed prairie of the Loess Hills and Plains is naturally divided into three communities. Blue grama, or blue grama with a small amount of buffalo grass, is a dominant in the short-grass type which occurred on hilltops and extended for some distance down the hillsides on dry slopes, sometimes to near the base of the hill. This type of prairie never occurred in ravines or on catsteps, where water content of soil was higher, or on steep banks. A postclimax type of tall and mid grasses of true prairie clothed the typically broad flat bottoms of ravines, low terraces, and lower slopes of hills. A third community of mixed short and taller grasses occupied the hillsides where the short grasses from the hilltops and upper slopes mingled with the mid and tall grasses from the lower land. This two-layered community is a typical expression of mixed prairie.

Almost no study of plant life in this region was made by botanists until 1946, despite the fact that by 1874 early settlers in the several counties located along the larger stream valleys where there was an abundance of fuel and water. By 1890 most of the land was homesteaded and the fencing of range land became general.

From 1946 to 1953 several extensive and detailed studies were made not only of the kind and distribution of the plant communities but also of their degeneration under grazing. By this time about half of the area had been placed under cultivation. Vegetation described here is that which occurred in ungrazed native mixed prairies which were mowed annually for hay.

FIG. 86.—A view in short-grass prairie. Blue grama is the chief species.

SHORT-GRASS TYPE

In this community two short grasses dominated, usually in almost pure stands or even where there was an intermixture of taller grasses, such as side-oats grama. Blue grama was far more abundant than buffalo grass. Since both of these grasses also occur widely on the plains of western Nebraska, they will be described briefly (Fig. 86).

Blue grama is a perennial warm-season short grass. In our latitude it does not produce new forage until about the middle of May but it develops rapidly and may produce flowers and seed within 60 days. It is a bunch grass of far greater drought resistance than buffalo grass or indeed most plains grasses. In drought it becomes dormant but revives quickly after rains. The mature foliage is only 3 to 5 inches tall but the numerous flower stalks attain heights of 6 to 20 inches. The bluish-green leaves vary from 2 to 6 inches in length. They are slender and often curled. The leaves are so close to the ground that even under intensive grazing much green, food-making tissue remains. This grass increases under heavy grazing where most mid grasses succumb. Often a second period of flowering occurs in the fall.

Buffalo grass is similar to blue grama in many ways. The most striking difference is its method of propagation by means of numerous runners or stolons which are sometimes 2 to 2.5 feet long. They root freely at the nodes or joints and produce additional plants, thus forming a sod. This warm-season short grass becomes green late in spring and dries early in fall. The foliage of this low-growing perennial is only 4 to 5 inches high and flower stalks are 4 to 8 inches tall. Pollen bearing and seed bearing flowers usually occur on separate plants, but a small percentage of plants have flowers of both sexes. The foliage, like that of blue grama, is palatable to all classes of stock but is produced in relatively small amounts. Both grasses furnish excellent grazing even after the plants have dried. This makes them valuable during periods of drought and for winter grazing. Because of their low stature and low productivity they are not important for hay. The houses of early settlers were made mostly from the tough sod of buffalo grass.

Although short above ground, these grasses are deeply rooted. Even in the hard lands of the Great Plain of Nebraska, Colorado, and Wyoming the roots of short grasses have an average depth of 4.5 feet. In the loess hills they are frequently 6 feet deep. Moreover, the roots and their branches are much finer than those of most grasses of true prairie. Although the roots of some grasses were confined to the upper 3 feet of soil, those of the most important ones and some of the forbs extended their roots to depths of 6 to 10 feet. Others were much deeper.

An extensive study of more than 30 prairies revealed that in this community blue grama alone usually composed about 90 percent of the vegetation and buffalo grass about 5 percent. Various sedges of low stature were widely scattered. Junegrass and western wheatgrass were present here and there but neither was abundant. Hairy chess and six-week fescue were weedy annuals, but this was chiefly a blue grama community. Usually only about a third of the soil surface was occupied by plants.

Of the forbs, lead plant, many flowered aster, and Missouri goldenrod of true prairie were fairly common, but certain species, such as scarlet globemallow, skeleton weed, and prairie coneflower, common on the Great Plains, were more abundant. This grassland type composed about a third of the mixed prairie.

Mid- and Tall-Grass Type

Mid and tall grasses, chiefly big bluestem, occurred on the steep banks of ravines, on catsteps, on well protected north and east hillsides, and on most lowlands. Indeed the broad flat lowlands are called hay pockets, since the yield is much greater here than elsewhere (Figs. 87, 88). These habitats are well protected from wind and in addition they receive much runoff water from the uplands. Of course, catsteps and their increased water supply may occur wherever the soil has slumped. Nodding wild-rye and switchgrass, which indicate the best water supply, also occurred in unusually wet places. But prairie cordgrass was absent since the loess soil does not retain the water sufficiently for this species. It does occur along flood plains of rivers where black clay soil has been deposited. Side-oats grama ranked second in importance among the grasses. It occurred widely and often in nearly pure stands on the catsteps and banks along the steep slopes of ravines or in patches in valleys. Junegrass and tall dropseed were common along with plains muhly, a species rarely found in true prairie. An understory of Kentucky bluegrass was usually an important part of lowland vegetation. Clearly this mixed mid- and tall-grass prairie, with the short grasses kept out by the shade, was a westward extension into the plains climate of many species of true prairie. This community occupied a slightly smaller area than that of the short grasses, perhaps only about 27 percent of the whole.

FIG. 87.—Community of mid and tall grasses in a flat-bottomed ravine. Chief species are big bluestem, Indiangrass and side-oats grama.

Fig. 88.—A native prairie mowed for hay. Prairie grasses remain undisturbed on the banks that are too steep for mowing. Some banks (left) are partly covered with shrubs.

Mid- and Short-Grass Type

This mixed prairie community usually consists of a mixture of mid and short grasses (Fig. 89). Little bluestem, before the great drought (1933 to 1940), and side-oats grama were the chief mid grasses, but purple three-awn and plains muhly (rare eastward), and June grass were also important mid grasses. This community occurred on the hillsides where the short grasses and low-growing sedges from the hilltops mingled with the mid grasses. They formed the characteristic layered vegetation of mixed prairie. The small amount of little bluestem following the great drought and its replacement temporarily by the rapidly spreading big bluestem should not confuse the reader.

This community occupies about a third of the entire prairie area. Western wheatgrass, a perennial, cool-season, mid grass is a great seeder and also possesses long vigorous rhizomes. During drought it invaded open land where other grasses weakened or died. At the end of the drought it was in almost complete possession of nearly 10 percent of the prairie. It will almost certainly be replaced slowly by the former occupants of the land as it was in the true prairie eastward.

FIG. 89.—Typical mixed prairie consisting of a lower layer of short grasses and an upper one of a mid grass.

About 24 principle species of forbs occurred in this general area. They included species both of true prairie and the Great Plains. They were most abundant in the mixed prairie community. Lead plant, a half-shrub and a legume which produced a large amount of palatable forage, was most abundant, conspicuous, and widely distributed (Fig. 90). Many-flower psoralea, also a legume, was among the most widely distributed of all of the forbs. About 50 other species were of secondary importance. Thus the total forb population was much smaller than that of true prairie.

Places dominated by shrubs were of limited extent and occurred only in the most moist habitats. These were on north and east-facing banks of ravines, in deep, narrow ravines, and in other locally protected places. Usually only thickets of rose bushes, lead plant, and wolfberry were found in more exposed places. Small thickets of dwarfed wild plum and western choke cherry occurred. They were more or less overrun by poison ivy and wild grape; currant and gooseberry often formed an understory. In well protected deeper ravines and along streams, box elder, American elm, red ash, and cottonwood occurred. Some of the trees were 40 to 50 feet tall and beneath them grew a dense, thicket-like, understory of shrubs. Buffalo-berry was sometimes present.

Dry weight of plants is one of the best quantitative characteris-

FIG. 90.—A society of lead plant among the grasses on a loess hill.

tics of vegetation. This, of course varies from year to year. In one study (1948, a good year) total yields were greatest, 1.7 tons per acre, in the mid- and tall-grass type. The maximum yields in certain hay pockets in the most mesic portion of the lowlands were 2.5 to 3.5 tons per acre. The short grass and mixed grass types each yielded about 1 ton per acre when cut by hand and close to the soil surface. In ordinary mowing about 40 percent of the short grass is left standing as stubble.

This extensive grassland has been used chiefly for grazing, largely for summer range, and especially for cattle. Consequently the vegetation has undergone various changes, somewhat in proportion to the intensity of grazing. Ranges vary in size from 80 acres to a square mile or more.

XI.

The Great Plains

The greatest expanse of grassland in North America occupies the relatively dry interior of the continent. In width it extends eastward from the Rocky Mountains across the Great Plains to the Central Lowlands and hence through eastern Colorado and Wyoming, halfway across Kansas and over all but the eastern part of Nebraska. This central part of the Great Plains is clothed with vegetation mostly of mid and short grasses which is known as mixed prairie (Fig. 91, 92). The true prairie, with its tall and mid grasses and great abundance of forbs, is characterized by a widely different climate than that of the mixed prairie of mid grasses of lesser height and an abundance of short grasses, which are only a few inches tall. Chief differences are greater precipitation, deeper and more constantly moist soils, and a greatly decreased rate of evaporation in tall and mid grass prairie.

Soils and Soil Water

The nature of the Brunizem and Chernozem soils has been discussed. The remainder of the state (Sand-Hill Region omitted for the present) is occupied by Northern Brown soils. They occur in an area of 21 inches or less rainfall. A line drawn from central Holt to Red Willow County indicates approximately their eastern boundary (Fig. 6).

With decreased precipitation the vegetation becomes sparser, the soil becomes lighter in color, and the solum thinner. The vegetation of this soil while not luxuriant is still well developed, especially the parts underground. The deeper layer of soil with less humus is lighter brown and contains the calcium accumulations. The lime zone decreases in depth with decreased precipitation, and the influence of the vegetation on the soil westward becomes less and less. The fertile dark brown soils are used for wheat production despite

FIG. 91.—Small tributary of the Republican River in southwestern Nebraska. Side-oats grama and sand dropseed are scattered through the understory of blue grama and buffalo grass (light areas). (From *Grasslands of the Great Plains.*)

the hazardous climate. Since the nutrients have not been leached, the soils, though low in organic matter, are fertile.

In a series of experiments begun in 1920 and extending over a period of three years, an attempt was made to employ plant production in ungrazed true prairie and mixed prairie as a measure of environment. Lincoln, in eastern Nebraska, was selected as representative of true prairie conditions. Phillipsburg, about 190 miles southwest of Lincoln (and due south of Holdrege) in north-central Kansas is representative of eastern mixed prairie; and Burlington, Colorado, about 180 miles farther west and south (and south of Chappell) represents a more central location in mixed prairie. The altitude of the several stations is 1,100 feet at Lincoln, 1,900 feet at Phillipsburg, and 4,160 feet at Burlington. Differences in elevation more than offset those of latitude; spring usually opens about 7–10 days later at Phillipsburg, and 18–23 days later at Burlington, than at Lincoln. Precipitation, a chief factor in determining the type of vegetation, varies from about 28 to 23 and 17 inches at the several stations, respectively, decreasing westward.

A glance at the distribution of the mean annual precipitation shows clearly that most of the moisture falls during the growing season and only about one-tenth during the three winter months (Fig. 93). The normal decrease of 5 inches at Phillipsburg under that at Lincoln, and a further decrease of 6 inches at Burlington are quite evenly distributed throughout the year.

Available water content of soil was ascertained at each station by soil sampling to a depth of 4 feet, at 12 periods from April to September. At Lincoln a sufficient amount of water was available to promote good growth. At Phillipsburg, July and early August were periods of drought and, at times, actual deficiency. Available water content at Burlington was favorable until June, but thereafter marked deficiencies were of frequent occurrence. The data are for 1920 but they are also representative of those of the two following years. Wind movement increased greatly proceeding westward. In fact, the conditions for plant growth as regards rainfall, available soil water, temperature, humidity, wind, and evaporation (all carefully measured) are normally most favorable at Lincoln, intermediate at Phillipsburg, and least favorable at Burlington. These conditions are indicated by the native vegetation and borne out by the growth of both native and crop plants.

Plant production of native vegetation was ascertained by employment of a large number of samples (each with an area of 10.8 square feet) at the three stations. This study was begun in 1920 and continued in 1921 and 1922 to ascertain the effects of soil moisture

FIG. 92.—Rocky hillside on breaks of the Cheyenne Table near Bushnell. The mixed prairie is composed largely of little bluestem, side-oats grama, Junegrass and needle-and-thread. (From *Grasslands of the Great Plains.*)

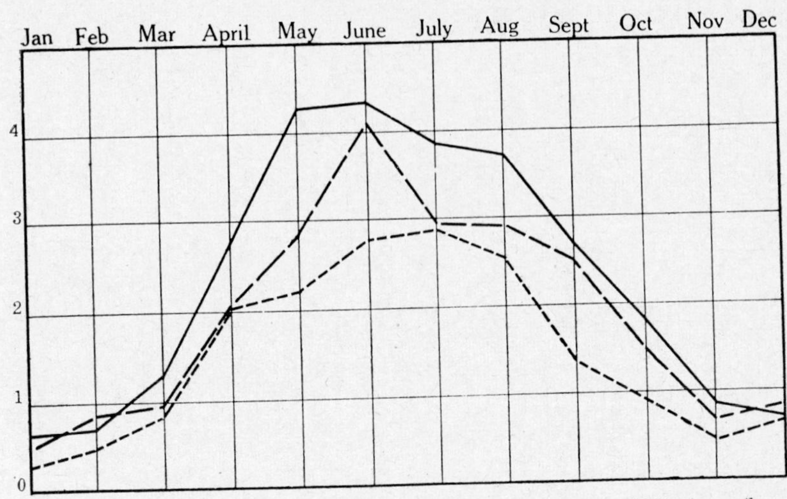

FIG. 93.—Mean annual precipitation at Lincoln (solid line), Phillipsburg (long broken lines), and Burlington (short broken line).

fluctuations on the yield from year to year. This gave an expression of the growth of each community as a unit. Since even the most uniform plant cover shows some variation in density, the samples were selected with much care and in sufficient number (about 50 at each station) to insure dependable results. Clippings were made, usually in late summer, and at approximately the same time at all stations. The clipped vegetation when dry was shipped to Lincoln, where it was uniformly air dried and plant production determined on the basis of dry weight.

Average dry weight of vegetation in tons per acre is somewhat higher than that of mowed prairie, since by hand clipping close to the soil surface the stubble is included.

STATION	1920	1921	1922
Lincoln	2.16	2.69	1.99
Phillipsburg	1.66	1.79	1.39
Burlington	0.80	1.50	1.00

Average production for each year at the several stations shows a graduated series, the yield increasing with amount and increased efficiency of rainfall (Fig. 94). The total yield at all stations was greater in 1921 than during the preceding or following year. That in 1922 was less at every station than in 1921. An examination of the record of available soil moisture revealed the cause of the difference. It was clearly determined that the water relations of soil and air (rate of evaporation) were controlling, other factors being merely contributory. Similar relations have been ascertained not only for the smaller cereals (oats, wheat, and barley) but also for alfalfa and sweet clover.

Climate

Climatic conditions over the mixed prairie are not easily described. The climate is one of extremes. It is commonly called semiarid, but in some years it is humid and in others desert-like. It is not a permanently established climate but a dynamic one with large scale fluctuations and wet and dry trends. The climate is subject to pronounced changes in temperature; hot summers with cool nights occur throughout. Seventy to 80 percent of the rain falls during the growing season, mostly in May and June. Much of the rain falls in heavy showers and is lost as runoff. Much rainfall is also lost in light showers followed by bright sunshine and wind, which promote rapid evaporation. Precipitation is frequently poorly distributed and drought strikes at unpredictable intervals. In mixed prairie drought is always imminent. Hot, dry, southerly winds of high velocities frequently parch the soil in summer. Hailstorms are common. Sometimes the native vegetation is beaten down, the tops of plants destroyed, and the soil pock-marked to a depth of half an inch or more. Often the range grasses cease growth in late July or early August, and not infrequently spring drought delays early development of the plant cover. Yet the wonderfully adjusted vegetation, to be described, has successfully met all these hazards through thousands of years.

The most significant difference of mixed prairie from true prairie is the almost universal presence of one or more short grasses or

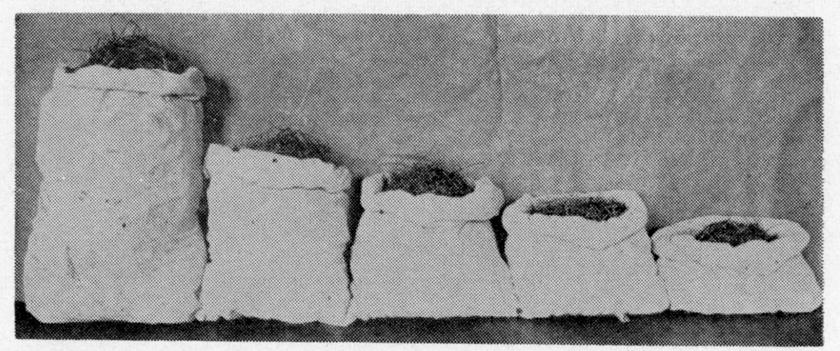

Fig. 94.—Average yield from 10.8 square feet of grassland. From left to right: Tall grasses from lowland prairie at Lincoln, mixed mid and tall grasses from upland prairie at Lincoln, mixed mid and short grasses from Phillipsburg, buffalo grass from Phillipsburg, buffalo grass from Burlington.

FIG. 95.—View in mixed prairie. Chief grasses are blue grama and an overstory of needle-and-thread.

sedges as a lower layer under the mid grass species. The distinctive feature of the mixed prairie is the intimate mixing of mid grasses with the shorter ones (Figs. 95, 96). The short, sod-forming grasses are chiefly buffalo grass and blue grama. They are well adapted to a brief growing season. For most true-prairie plants the Great Plains is an uncongenial habitat because of the high evaporating power of the air, and especially the shortage of water during late summer, just at a time, often, when their demands upon the habitat are greatest.

Competition between the two types of vegetation seems nicely adjusted. While the short grasses have a disadvantageous light relation because of their height, this is counterbalanced by their more favorable position regarding evaporating power of the air and especially their greater ability to withstand grazing. Eastward, under more favorable conditions for growth, the short-grass layer disappears as the mixed prairie merges into true prairie.

VEGETATION

The Loess Hills and all of Nebraska west of true prairie are a part of the Great Plains; the type of vegetation is mixed prairie. The mixed prairie clothes the most extensive grazing area in North America. This great grassland derives its name from the fact that the climax or original plant cover, undisturbed by civilized man, was composed of both mid and short grasses intermixed and occurring on more or less equal terms. Although the vegetation is not uniform everywhere, many of the more abundant and important grasses occur throughout. For example needle-and-thread, western wheatgrass, and Junegrass continue across the plains and into the foothills from Montana through Wyoming and Colorado. Blue grama is plentiful from Saskatchewan to Texas. In one study, an experienced investigator found one short grass and at least one mid grass together as dominants in all but 15 of 393 localities examined.

In a study of the history of range use, it has been stated that since the coming of the herds of the cattlemen and the settlement of the land, the story of the Great Plains is one of prodigal exploitation of a vast natural resource on a scale never witnessed before. This began in 1880–85 and was aggravated from time to time by severe droughts, while settlement continuously decreased the amount of range land.

Wet seasons on the Great Plains, by increasing water content and decreasing competition, regularly reproduced mixed prairie in a convincing manner during the first quarter of the present century, over wide areas that were apparently covered with short grass during dry years. The reader is referred to *Grasslands of the Great*

FIG. 96.—Prairie of needle-and-thread and western wheatgrass with some blue grama outside the fence and ungrazed. The mid grasses have almost disappeared in the pasture. (From *Grasslands of the Great Plains.*)

FIG. 97.—Typical western range on hardland near Scotts Bluff.

Plains for further information and numerous studies by many investigators (Figs. 97, 98).

The manner in which tall and mid grasses of true prairie gradually give way to mid and short grasses westward has been described. The way in which short grasses, especially blue grama, clothed the hilltops and upper slopes of the loess hills and also intermingled with mid grasses to form the mid- and short-grass type has also been explained. Sandy soils in the Great Plains support a quite different type of vegetation, mostly because water penetrates more readily than in other soils which are called hardlands (Figs. 99, 100). Box Butte, Cheyenne, and Perkins Table as well as much of the Dissected Plains (Fig. 7) are all examples of hardland where runoff is high.

Mid Grasses

Mid grasses are circumpolar in origin. The short grasses and side-oats grama came in from the southwest, and the bluestems and other tall grasses are, as a rule, southern or subtropical. Conversely, the several species of grass-like sedges are of northern derivation. The intermixing of these species derived from so many sources is another reason for the name "mixed prairie." Important variations

FIG. 98.—Detail of short grass range near Wauneta.

FIG. 99.—The western range closely grazed, with clumps of cactus. Photo near Crawford.

FIG. 100.—Blue grama and thread-leaf sedge heavily grazed on a range near Scotts Bluff.

in mixed prairie are due to the distribution of the rather numerous dominants, especially from north to south. Some dominants are found wholly or largely in the south. Curly mesquite and galleta are examples.

The chief mid grasses are side-oats grama, needle-and-thread, western wheatgrass, sand dropseed, Junegrass and green needlegrass. In addition, little bluestem, and even the tall grass, big bluestem, with certain other species, form distinct types of postclimax (beyond their climate) communities where topography and soils are such that they benefit from an unusually large water supply.

Side-oats grama is a very abundant and important species of mixed prairie. It is a perennial, warm-season plant with short rhizomes and is most abundant on rough soil of breaks and on lighter soils. This early growing grass reaches a height of 1.5 to 2 feet, and remains green late in fall. Although not so drought resistant as blue grama, it reseeds readily and furnishes excellent grazing and good hay. In Figure 61 other characteristics may be noted, especially in regard to flowering.

Needle-and-thread is a perennial grass which becomes green late in March or early in April. It forms tufts or bunches, not closely spaced, attains a height of 1 to about 3 feet and flowers early in June (Fig. 101). Because of its early growth and high nutrient qualities, it is apt to be grazed out. It reaches maturity and ripens its seeds (fruits) in July. These are sharp-pointed and often injure sheep, especially, and other grazing animals. When used for hay, mowing should be done either before the seeds are ripe or after they have fallen. The name is derived from the long-awned seed which resembles a threaded sewing needle.

Western wheatgrass is a hardy perennial which begins growth in March and develops rapidly (Fig. 102). Owing to its many long, much-branched rhizomes, it spreads widely to form a dense sod. It often grows in dense, nearly pure stands, but it is usually intermixed with blue grama or other grasses. A height of 2 to 3 feet is attained in good years but it is much dwarfed during drought and may not flower. It is a very leafy plant with a bluish-green color. The spikes, which are 2 to 6 inches long, are prominent above the foliage level. Blossoming occurs in June. This grass is often intermixed on uplands with needle-and-thread, thread-leaf sedge and

FIG. 101.—Needle-and-thread and Junegrass. (From *Grasslands of the Great Plains*.)

FIG. 102.—Sand dropseed and western wheatgrass. (From *Grasslands of the Great Plains.*)

blue grama. Wheatgrass furnishes excellent forage for cattle and horses the year around. Like most mixed prairie grasses, it remains palatable on drying and retains much of its nutrients.

Sand dropseed is a drought-resistant grass which forms tufts or open bunches (Fig. 102). It is most abundant on dry, coarse soils and in sandy areas. The stems, which often spread outward at the base, attain a height, when mature, of 1 to 3 feet, but the foliage is mostly confined to the lower half of the plant. Flowering occurs in late summer and fall. The tiny seeds are extremely abundant, occurring by the thousands in a single panicle. They drop to the ground and are scattered as dust by the wind. This grass is not so highly relished by livestock as wheatgrass, but it does produce a fairly large amount of palatable forage, especially in early summer.

Green needlegrass is another mid grass of wide distribution in mixed prairie, especially northward. Its best development occurs in swales or on flat land that is occasionally flooded. It is a perennial bunch grass with a height varying from 1 to 3 feet and with mostly basal leaves. Growth begins early in spring and flowering about the first week in June. Growth may continue until autumn, except during drought. There is an inch-long, bent awn on each seed. The awns are much less troublesome to livestock than are those of needle-and-thread. This grass furnishes good pasture for all kinds of livestock both summer and winter, and may be grazed at any time without injury to the animals.

Junegrass ranges widely over prairie and plains. It has been described in true prairie, but it is of smaller stature on the plains. It is neither long-lived nor very drought resistant but maintains its stand by rapid reseeding (Fig. 101).

The preceding grasses are the most important and widely distributed; there are, of course, several other mid grasses on the Great Plains. Blue grama and buffalo grass are the most abundant of all the grasses or grass-like sedges of the short grass group. Hairy grama also occurs on rough, rocky ridges and on lighter types of soil.

Short Grasses and Sedges

A brief description of buffalo grass and blue grama has been given. The characteristic life form of mixed prairie is perennial grasses and grass-like, dry land sedges. Where weedy grasses such as tumblegrass and downy chess occur in abundance, it is a distinct sign of some sort of disturbance.

Red and purple three-awn are perennial bunch grasses which were formerly called wire-grasses. The bunches of red three-awn are 3 to 6 inches wide and 7 to 12 inches tall. The stems are fine and unbranched. The very narrow leaves are mostly basal with a harsh upper surface. They are inrolled, sharp-pointed, and contain much fibrous tissue. The spikelets have a reddish tinge but the plant as a whole is conspicuous because of its silvery gray color. The seeds are sharp-pointed and each of the three awns is 3 inches or more in length. The abundant seeds are scattered far and wide by the wind as well as by animals. Purple three-awn forms more upright and taller bunches, usually 12 to 20 inches high. The stems are coarser and not so densely aggregated. The awns are shorter (1 to 2 inches) and usually the panicles are purplish in color. Both species grow best on somewhat sandy soils but are widely distributed except in low wet places. They flourish in bared or disturbed soil. In both Nebraska and Kansas these grasses are usually considered worthless for forage. Indeed the hard, sharp points of the seeds often injure eyes, nose, and mouth of grazing animals and damage the hide of sheep.

Thread-leaf sedge is grass-like both in appearance and habit. It is a perennial with a very long life span and grows in dense bunches to form a tough sod (Fig. 103). The light green, threadlike, tufted leaves produced by each bit of sod are 4 to 10 inches long in wet years but otherwise only 2 to 5 inches in length. The slender flower stalks, usually a little higher than the leaves, produce a brown, chaffy spike or flower head only about an inch long. This plant, of northern origin, resumes growth somewhat earlier than western wheatgrass and a month earlier than blue grama. Thus it is especially valuable for grazing in very early spring. Growth is completed late in May, but this sedge maintains its high palatability for all classes of stock until about the first of July, when it becomes dry and tough. In Nebraska it is most abundant north of the Platte.

Needle-leaf sedge is another species which ranks high among the plants of mixed prairie. It has smooth, stiff, pale-green stems which form small tufts. They originate from rhizomes and result in dense patches which furnish very early grazing. There are various other sedges intermixed with the grasses and forbs, but none so important as these in midwestern grazing lands. Penn sedge, for example, is a low-growing perennial which ranges widely in both true and mixed

FIG. 103.—Upland mixed prairie in climax condition. Prominent bunches in foreground are thread-leaf sedge. Western wheatgrass is the chief mid grass; general understory is blue grama and buffalo grass.

prairie. It resumes growth and flowers very early in spring. Because of its abundant rhizomes it often grows in patches.

An extensive account of the grasses and grass-like plants from Montana to Texas is given by Weaver and Albertson in their book on *Grasslands of the Great Plains, their Nature and Use.* On many ranges the mid grasses are nearly grazed out and then the short grasses appear to be the dominants.

Drought resistance is an extremely important characteristic of range grasses. Near the end of the long drought cycle of the 1930's, in the summer of 1939, a survey of conditions was made by Dr. F. W. Albertson and the writer. It included 88 ranges selected as representative of grazing lands of western Kansas and Nebraska, portions of southwestern South Dakota, eastern Wyoming and Colorado, and the Panhandle of Oklahoma.

Severe drought, overgrazing, burial by dust, and damage by grasshoppers had resulted in greatly reducing the cover of range grasses. This part of the mixed prairie had almost completely lost its upper story of mid grasses on the non-sandy lands. The short grasses and sedges had undergone a process of thinning which had resulted in only the most vigorous plants remaining alive. The basal cover of grasses, normally about 80 percent, was greatly reduced. Only 16 percent of the ranges had a cover of 21 percent or more. In another 16 percent, cover varied between 11 and 20 percent. It was only 6 to 10 percent in 28 percent of the pastures and was reduced to 2 to 5 percent in another group totaling 16 percent. The remaining pastures, 24 percent, had a cover of only 1 percent or less.

Extremely poor conditions varied with the better ones throughout. The bare soil, during periods with moisture, was populated with annual weeds, chief of which was Russian thistle. In many ranges it was only with difficulty that one could distinguish denuded pastures from weedy, tilled land. It was observed that blue grama and not buffalo grass endures the drought best.

A very extensive study of the damaging drought cycle of 1952 to 1955 on the Great Plains was made by Dr. F. W. Albertson and two other competent range men. This research extended from New Mexico through western Kansas and eastern Colorado into Wyoming, South Dakota, and western Nebraska.

Among 24 representative ranges in the various parts of the Great Plains, losses of vegetation ranged from 10 to 99 percent. The heavier losses were nearly all in ranges with a high degree of utilization. They state that grasslands, weakened by overgrazing during wet cycles, are extremely sensitive to deficient soil moisture when drought strikes. Loss on heavily grazed ranges often was nearly double that of moderately grazed ranges and frequently more than double the amount on the nongrazed grasslands.

It is pointed out that the pioneers observed a higher carrying capacity of range land during wet periods and great decreases in drought, but the contribution made by overgrazing to drought losses was largely overlooked. Presumably native vegetation developed under conditions similar to those of today, and it is also safe to assume that native plants will continue to dominate the prairies if not continually overgrazed by livestock or buried too deeply by soil blown from cultivated fields. Therefore, if our native vegetation is completely destroyed, man should be held accountable.

Root Systems of Grasses

Extensive examinations of the roots of native grasses of mixed prairie have been made. Buffalo grass was excavated at eight widely scattered stations in mixed prairie and also at Lincoln (Fig. 50). Depth of the root system varied from about 4.5 feet at Limon, Colorado, to about 7 feet at Yuma, Colorado, and Ardmore, South Dakota. Several examinations at Lincoln revealed a depth of about 6 feet. At all stations except in true prairie, the fine roots spread laterally about a foot on all sides of the clump at a very shallow depth. The surprising depth of this grass indicates that, throughout this area of Great Plains, moisture is available for growth, at least periodically, in the deeper soil layers.

Blue grama was examined at four widely scattered stations on the Great Plains and at Lincoln. Little difference in root distribution was found in the several plant communities, except that the marked development of widely spreading surface laterals so common in the more arid portions of the grassland did not occur in true prairie. Thus, both buffalo grass and blue grama are short only above ground. The roots extend deeply.

Roots of western wheatgrass were examined at five stations in mixed prairie. Depth of root penetration varied from 5.2 feet at Limon to 7 feet at Colorado Springs, in Colorado, and at Ardmore, South Dakota. At Lincoln the depth was 8 to 9 feet. Little bluestem is usually 5 to 5.5 feet deep at Lincoln. A similar depth but somewhat wider spread of roots in the surface soil was ascertained at both Limon, Colorado, and Phillipsburg, Kansas. Depth of root penetration of other grasses and sedges was about the same (5 feet) as that of upland grasses in true prairie.

The development of roots and tops of four abundant range

grasses is shown in Figure 104. The rhizome habit, so pronounced in lowland of true prairie, and the bunch type of upland true prairie are both represented by species of the Great Plains. In addition, buffalo grass exhibits the presence of stolons. The roots, of course, compete for water and nutrients in both prairie and plains; in the figure the plants are set apart for the sake of clarity. The fact that needs most emphasis is that the short grasses are short only above ground. Their root to top ratio is much greater than that of mid grasses.

FIG. 104.—Tops and roots of dominant grasses common to mixed prairie. Only those parts in a strip one inch wide are shown. From left to right are buffalo grass, purple three-awn, western wheatgrass and side-oats grama.

The roots of green needlegrass growing in a swale have been traced to a depth of 11 feet. Roots of sedges are also quite deep. The plants often grow in clumps 4 to 6 inches in width. They are furnished with an enormous number of tough, wiry roots which seldom

FIG. 105 (upper).—Western wallflower, a mustard, and prairie cone flower.

FIG. 106 (lower).—Two stems of milk vetch with yellowish white flower, and purple prairie clover. (Photos from *Grasslands of the Great Plains.*)

descend vertically but run obliquely away from as well as under the plant, forming a great tangled mat to a depth of 1 to 1.5 feet. The horizontal spread is often 2.5 feet in the soil surface. Lateral roots end in brush-like mats of branches. Many roots were 4 feet deep and not a few exceeded 5 feet.

Roots of needle-and-thread extend deeply in the hard, dry plains soil. A lateral spread of 2 feet is usual and depths of 5 feet are attained. The only shallowly rooted perennial grass was Junegrass. It has been examined in many types of soil from eastern to western Nebraska but always with the same result, a maximum depth of about 2 to 2.5 feet.

Forbs

Mixed prairie may be considered from many points of view; a most important one is that of the species of which it is composed (Figs. 105, 106). Although the kind of prairie is distinguished by its dominant species, subdominant and secondary species are also important. Forbs are always present and often abundant in prairie. Often they are more conspicuous than the grasses but they are nearly always less important. The number of perennial forbs in mixed prairie is also large but smaller than in true prairie. Moreover, the abundance of the individuals is reduced because of decreased precipitation and drier soil. Some species are rare, others are common, and still others occur abundantly. By far the greater number of species are composites with flowers grouped in heads (such as gumweed and sunflower) and legumes. Other families furnish only a small percentage. While a few plants are poisonous, most are valuable as forage. The list of mixed prairie forbs that are regularly eaten by livestock is a long one. They provide a valuable variety in the diet of grazing animals and are especially rich in calcium and phosphorus. Cattle do better on mixed herbage than on grasses alone. A list of the important forbs includes the following:

Composites	Legumes	Others
Blazing star	Crazyweed	Cactus, species
Broom snakeweed	Drummond vetch	Narrow-leaved four o'clock
Cut-leaved goldenweed	Few-flowered psoralea	Narrow-leaved puccoon
False boneset	Ground plum	Plains milkweed
Fringed sage	Lead plant	Prairie rose
Gumweed	Missouri milk vetch	Prickly poppy
Hairy golden aster	Purple prairie clover	Red false mallow
Many-flowered aster	Silver-leaf psoralea	Scarlet gaura
Missouri goldenrod	White prairie clover	Small soapweed
Prairie coneflower	Woolly loco	Tooth-leaved primrose
Purple coneflower		Western wallflower
Stiff goldenrod		Wild onion
Wavy-leaved thistle	(Figures 107, 108)	

FIG. 107 (upper).—Wavy-leaved thistle, and prickly poppy with large white flowers.

FIG. 108 (lower).—Purple coneflower and broom snakeweed. (Photos from *Grasslands of the Great Plains.*)

Root Systems of Forbs

To what depths and how thoroughly do the forbs of the hardlands occupy the soil? Careful study of many root systems has been made in several places and the roots measured and drawn to scale. To save much space 12 species are shown in one group in an excavation 20 feet deep (Fig. 109). For general depth of roots compared with those in true prairie, refer to Figure 78.

Although some roots were quite extensive, the average depth of penetration was only 6.8 feet. This is 5.7 feet less than that of the group from true prairie. Moreover, the more thorough occupancy of the surface soil than in true prairie is remarkable. This may best be realized by inverting the figures. Cacti, as illustrated by number 8 on Figure 109, spread most of their roots widely, often 5 to 6 feet on all sides of the plant, and the roots branch profusely. Only a very few roots extend downward. Several species were examined. Plants number 3, 7, and 10 (fringed sage, broom snakeweed, and ragwort) extend their roots far outward in the surface soil be-

FIG. 109.—Excavation 20 feet deep showing root habits of several forbs of mixed prairie hardlands. 1, gromwell; 2, four-o'clock; 3, fringed sage; 4, hairy golden aster; 5, whorled milkweed; 6, prickly poppy; 7, broom snakeweed; 8, pricklypear cactus; 9, Lambert crazyweed; 10, ragwort; 11, scurfpea; 12, rush-like lygodesmia or skeleton weed.

fore turning downward. This may be seen best by again inverting the figure. Thus they are extremely competitive for water with the grasses and sedges, which, for clarity, were left out of the picture.

Small soapweed also has widely distributed lateral roots, many 20 to 30 feet long. Beginning at the soil surface and to a depth of 6 or more feet, coarse, poorly branched fibrous roots occur in great abundance. They absorb water and nutrients from an enormous volume of soil. Conversely, other root systems, such as prickly poppy (number 6) and skeleton weed (number 12) are often almost unbranched. Coarse laterals, however, depend only little upon surface soil moisture but may penetrate to depths of 12 or more feet.

Among 36 species of plants examined in the hardlands of Nebraska and several western adjoining states only 4 were entirely rooted in the surface 2 feet of soil. Sixteen species reached depths of 5 feet. The remaining 16 species extended their roots below 5 feet and many to a depth of 7 to 9 feet. None were found deeper than 13 feet.

This description of mixed prairie as it appeared in the first quarter of the present century is very different from the studies made during the extremely severe drought years of 1933 to 1940. The vegetation waxes and wanes and a single description at any one time cannot tell the whole story. Indeed typical native mixed prairie, like that of bluestem prairie, has almost disappeared. This has resulted from breaking the sod or destroying nearly all of the native vegetation, except the most persistent species, by more than a century of fencing and overgrazing.

XII.
Sand Hills

Nebraska has the largest area of sandy soils and sand dunes of any of the plains states. Physiographically, the loose sandy soil is its chief characteristic, although the hills and valleys are more sharply defined than are those of true prairie. An area of dune topography of more than 18,000 square miles occurs in North Central Nebraska (Fig. 7). The western border has an altitude of about 4,000 feet, but on its eastern margin the elevation above sea level is only 2,000 feet. Because of the unstable sand and strong winds, this sand dune area has a type of vegetation quite unlike that of the true prairie eastward or the hardlands mixed prairie (Figs. 3, 110, 111).

A study of the vegetation by experienced investigators began in 1889 and several major studies on the native flora both above and below ground have been continued by botanists, plant ecologists, and others to the present time. It is the plan of the writer to describe the vegetation as it appeared in the early part of this century, although various changes have occurred since that time.

An early description of the sand hills follows.

> The hills are all round-topped or conical and smooth, clearly showing that they have been shaped by the wind. There are many depressions between hills, many of which assume the proportions of valleys more than a mile in width and sometimes many miles in length. From these well developed valleys the low places decrease in width and length until they are mere narrow, saucer-shaped basins or "pockets" a few hundred yards across. The well-pronounced valleys are, as a rule, about parallel and trend in a southeast and northwest direction. Such valleys are frequently enclosed by ranges of hills and in this way effectively separated from adjacent valleys, though such may not be more than a half mile distant. Sometimes instead of the valleys being separated by round-topped hills this is accomplished by a continuous rounded ridge. The sides of these hills are often very steep, making difficult the direct passage from one valley to another. In the regions characterized by short valleys and basins the general

landscape is strikingly different because in such places the hills rise on all sides without any regularity. Low hills, intermediate hills and high hills are closely associated, with no long separating valleys. The result is a very abruptly rolling

Fig. 110 (upper).—Sand hills with a good cover of needle-and-thread. Photo Dr. R. J. Pool.

Fig. 111 (lower).—View in the sand hills about 1900, showing the open nature of the cover and the dune topography.

> surface with rounded or oblong depressions of varying depth, with the rounded or conical dunes above. Height of the hills varies from a few feet to 150 feet or more (Figs. 112, 113). A single hill may occupy only a few acres of land but a large one may extend over a square mile in area.*

Other investigators (1900) report that in general the most noticeable character of sand-hill vegetation, after one has become accustomed to the great variety of species which the sparse vegetation of each hill affords, is its extreme monotony. This is due to the predominance of bunch grasses, which are the controlling element in the covering of the hills, hillsides, sand ridges, and sandy tablelands of the watershed. These bunches, which are the most characteristic feature of the sand hills, enable a large number of other species to occupy the intervals sparsely. But in many large areas, where the bunch grass had no foothold, or where the shiftings of the sand have at length dried it out and caused the hills to become barren, different conditions prevail and the sand may be blown away. The sand hills and sandy ridges through the region present an open formation which shows almost no change in any part. Not only has every hill the same species as every other hill,

Fig. 112.—General view of the sand dune topography; the east end of Dewey Lake is in the background. Photo W. L. Tolstead.

*R. J. Pool, "Glimpses of the Great American Desert," *The Popular Science Monthly,* 80 (1912), 209–235.

FIG. 113.—A large cattle range in the Sand Hills. Abundance of sage indicates overgrazing.

but they occur in the same manner and have the same relation, and in consequence one hill, as a rule, has the same appearance as another. The vegetation of the sand hills is sparse, notwithstanding that more than 200 species occur; plants are widely spaced; and sand is everywhere to be seen. Very rarely do certain plants grow densely or even closely together as in other prairies, but the individual plants or tufts of plants are commonly 8 to 30 inches apart.

From a physiographic standpoint the Sand-Hill Region of Nebraska is a plains region that has been excessively modified by the force of the wind. It is a region of changed environment caused by wind. There is little surface runoff but there are many exposed groundwater lakes and marshes. The few streams are in valleys with alluvium and are fed by groundwater underflow. Most of the sediments forming the hills were blown from the easily eroded underlying sandy formations by strong northwest winds. The hills were mostly formed during the glacial period.

Because of wind, little soil formation has occurred. The hills are composed mostly of fine grained sand of a light yellow or pale brown color. Some of the tops of dunes have a coarser texture of sand than the sides and adjacent dry meadows. In the dry meadows small amounts of organic matter have accumulated in the surface few inches. But where tall grasses grow in the wet meadows there is a considerable amount of organic matter. The sand has a low water-holding capacity, especially the coarser sands.

The climate is similar to that described for mixed prairie. The prevailing winds in winter are from the west and northwest. Greater wind erosion occurs in winter. Blowouts usually develop on the northwest side of dunes. Although the sides of a young blowout are steep, as erosion continues the dune permits movement of wind over an elevation with a minimum of friction. Plants cannot revegetate a dune until this condition has occurred, that is, when the blowout has matured.

The climate of the sand hills is the same as that of adjacent hardlands. Precipitation is about 23 inches near the eastern border but only 18 inches northwestward where the sand hills give way to hardlands. Most of the rain falls from April to June, inclusive, and late summer drought is common. Rainfall is absorbed with little or no runoff. Much of it reaches the water table. Soon after a rain the surface soil becomes dry and has a great retarding effect upon further evaporation. At a depth of a few inches the sand is always moist. There is a water table deep beneath the soil surface. Streams have a continuous flow due to drainage from the ground water. Springs are plentiful. Tall grasses can thrive on the dunes because the sand is efficient in absorbing rainfall without loss by runoff and in almost preventing evaporation from its surface.

Wind is one of the most important environmental factors affecting sand-hill vegetation. Summer winds are mostly from the south. In winter, winds of 40 to 60 miles per hour are not uncommon. Often wind velocity is high during several consecutive days and much sand is removed (Figs. 114, 115). Plants cannot revegetate a denuded area where wind erosion is severe. Blowouts are one of the most characteristic features of the sand hills. Wherever the sand is not held together by the roots of plants or by moisture, or is not otherwise protected, it is little by little carried away by the wind. If a spot on a dry hill becomes bare, the loose sand is blown away; a hollow is made; the surrounding grass dies from drought; the dry sand, no longer held together by the roots, slides down into the hollow and in its turn is borne away; and thus the hollow becomes larger and larger.

In the beginning, blowouts cover only a few square yards and are only a foot or less in depth, but in extreme cases they become hollowed out to a depth probably exceeding 100 feet and the circumference may be more than 600 feet. On the windward side a

FIG. 114.—A large active blowout. The prominent ridge in the foreground is the rim of another blowout. Photo R. J. Pool about 1913.

blowout is an area of sand erosion, but on the leeward side of the rim there is an area of sand deposit. The coarser sands are deposited first, but the finer particles are blown to varying distances beyond. At first both areas are places of disturbance, but finally they are stabilized.

COMMUNITIES OF THE SAND HILLS

The plant communities are distributed largely according to the amount of available water. Of the numerous lakes that occur in the Sand-Hill Region many are small; others are sometimes a mile in width and 1 to 5 miles long. A few that are undrained are saline and have little vegetation in or about them. But most are shallow, fresh water lakes. The populations of submerged and floating plants and bulrushes, cattails, and wild rice, which grow in zones to the edge of the wet meadow, are similar to those described elsewhere.

Wet Meadows

The annually flooded zone at the edge of the wet meadow supports numerous species of sedges, bluejoint, and some other hydric grasses as well as swamp milkweed, water hemlock, and other wetland forbs. Prairie cordgrass occurs in the wetter part of the adjoining meadow, but the most common grasses are big bluestem, switchgrass, Indian grass and slender wheatgrass. The water table in this zone is high in spring and early summer (1 to 2 feet below the soil surface) but otherwise 3 to 5 feet deep. This vegetation consists largely of grasses and forbs that have extended far westward from the true prairie as they did in the lowlands of the Prairie Plains.

Between the big bluestem and the lower slopes of the hills, upland true prairie occurs. In this part of the meadow land the water table has a minimum depth of 4.5 to 6.5 feet. Only the more xerophytic species occur. Dominant grasses are little bluestem, tall dropseed, Junegrass, and western wheatgrass. Most common true prairie forbs are stiff sunflower and Missouri goldenrod. On the slopes and tops of the dunes, quite beyond the reach of the water table, typical sand-hill vegetation prevails. These are the conditions found by Dr. W. L. Tolstead in the northern sand hills where extensive research was carried on. These low-lying grasslands are locally designated as hay meadows. In addition to the hay meadows, other wet meadows elsewhere were dominated by communities of ferns, or water hemlock.

Fig. 115.—North slopes of low sand hills near Dewey Lake. The shelters contain weather bureau instruments; the vegetation has been grazed but little for several years. Photo W. L. Tolstead.

FIG. 116.—Blowout grass in bloom in its natural habitat of loose, windblown sand. (From *Grasslands of the Great Plains*.)

Blowout Community

Blowout grass is the first species to become established in the bottom or on the lower slopes of the blowouts. But this occurs only after the wind becomes ineffective in further increasing its depth. It is a tall mesophytic grass which forms a very open growth (Figs. 116, 117). This habitat is one of strong winds, high temperatures, and bright light, but evaporation from the sand is not great and usually there is a water supply throughout the entire growing season.

The usually sparse and rather small clumps of blowout grass are connected by coarse, tough rhizomes which are frequently 20 to 40 feet long, as one may plainly see as they are exposed on the soil surface where the sand has been blown away. Often the course of the rhizomes in areas of deposit is vertically upward toward the new surface of the sand. A vast network of extremely well branched roots develop from the rhizomes. They spread horizontally through the surface sand and some grow vertically downward to depths of 7 or more feet. Thus this grass is unlike any other sand-hill grass in its underground habits.

Blowout grass grows on both the windward and leeward sides of dunes and in other disturbed situations. It is not much grazed where other forage is present but is eaten by livestock in autumn since it remains green after most other grasses have matured and dried. The areas suited to blowout grass have been greatly reduced since the stabilization of sand-hill habitats by control of fire and development of vegetation under grazing, which removes much of the inflammable forage.

Blowout grass is a pioneer and is rarely found on stabilized soil. When a thin stand is developed in the bottom of a blowout a few other grasses may appear. Chief among these are sand lovegrass, Indian rice grass, needle-and-thread, and sand reed. Moreover, several species of ordinary sand-hills forbs commonly appear in the blowouts after the blowout grass has somewhat stabilized the cover of sand. They are well adapted underground for growth here and in the region generally. After extended observation in hundreds of blowouts, early investigators found these to be tooth-leaved primrose, spiderwort, annual umbrella plant, rattlepod, white flowered spurge, and woolly, yellow hymenopappus. The spiderwort has a somewhat fleshy root system which spreads widely on all sides of the plant in the surface foot of soil. Roots of the annual umbrella

FIG. 117.—Bunches of sandhill muhly colonizing the bare sand.

plant have an even wider surface spread in addition to a five-foot taproot. Toothed-leaved primrose spreads widely in the surface and the underground parts give rise to numerous new plants. Roots are abundant to a depth of 4 to 5 feet.

A legume, lance-leaved psoralea, thrives in blowouts but it is also widely distributed elsewhere (Figs. 118, 119). It is a much branched upright plant 1 to 2 feet tall, with bluish-white flowers. A deep taproot (shown later) branches widely and may attain a depth of 9 feet. The individual plants, which are often several feet apart, are usually connected by rhizomes which may originate at various depths and spread outward to distances of 10 to 30 feet or more. Root tubercles occur at great depths.

Sandhill muhly is also a pioneer but is on relatively stable dune sands and usually acts as a successor to blowout grass. It has a more compact growth and a very efficient root system. Much attention has been given to the underground parts since the ecological relations of the dominant grasses and forbs were understood only after the roots were studied. This grass is characterized by tufted stems

FIG. 118.—Lance-leaved psoralea, a legume that is common in the sandhills, especially in blowouts, but it also thrives in dry meadows.

Fig. 119.—Hairy prairie clover. With its dense cover of silvery leaflets, it stands out as a prominent object in the sand-hills flora.

and glaucous, narrow, rigid leaves (Fig. 117). The culms, which arise from rootstocks, are grouped into small tufts or cushions that lie close to the sand. Clusters of roots arise from the rootstocks, which are only 2 to 6 inches long. While some of the tough, wire-like roots penetrate rather vertically downward to a maximum depth of about 3 feet, others run off obliquely at various angles or parallel with the soil surface, spreading 1 to 2 feet on all sides of a large clump. All parts of the roots are furnished with multitudes of very fine absorbing laterals.

Sandhill muhly is often abundant on sandy waste lands resulting either from fire or excessive grazing and trampling. The stiff sharp-pointed leaves are not readily grazed, hence it remains in badly overgrazed ranges. The reddish inflorescences are held 10 to 14 inches high and quite above the prickly foliage.

Bunch Grass Community

Bunch grasses form the most characteristic vegetation of the sand hills (Figs. 120, 121). The most important species at the turn of the century were little bluestem, sand bluestem, sand reed, and needle-and-thread (Fig. 122). This is the same little bluestem that is so abundant in true prairie eastward. According to all investiga-

FIG. 120.—Sand lovegrass and sand reed. (From *Grasslands of the Great Plains.*)

tors before the great drought of 1933–40, it was the most frequent and abundant of the grasses in the sand-hills landscape. But its losses by drought, which were 90 to 100 percent, equalled or exceeded those in true prairie. Mature bunches varied from less than 1 to 2.5 feet in width. From one-half to two-thirds of the exceedingly well developed root system spread laterally in the surface 1.5 to 2 feet of sand (Fig. 122). Since it was scarcely grazed by livestock in the sand hills (although relished in true prairie) but furnished abundant fuel for fire, cattlemen were not much concerned about its losses in drought.

Sand bluestem is shown in Figure 121. It is a coarse grass 3 to 5 feet high with only a few stems in the loose, open bunches. It maintains an open growth even in mature communities. Like big bluestem of the true prairie it has a high forage value and today is far less abundant than 60 years ago. The relatively short rhizomes and abundant, deep roots are shown in Figure 122. Often 6 to 9 of the short branches occur on a single inch of root.

Sand reed has stems 2.5 to 4 feet tall which mostly arise singly from strong rhizomes. Rhizome propagation is pronounced. Sometimes the wiry, much-branched, and matted rhizomes are intermixed with the roots to a depth of 2.5 feet as a result of drifting sand. They form a network connecting widely spaced plants. This grass now occupies so much of the area lost by little bluestem that it became the most characteristic grass of the sand hills. Well developed stems have 10 to 12 long leaves. It is drought resistant and occurs generally over the hills. This grass alone furnishes nearly half of the total forage in most places. The roots are tough and wire-like and they often spread widely. They fill the soil with an extensive meshwork and help to hold it firmly in place.

Sand lovegrass is also a perennial bunch grass of much importance (Fig. 120). The bunches are widely spaced and the plants are 2 to 3 feet tall. It grows in places on the north side of dunes which are protected from the dry south winds of summer but is found

Fig. 121.—Sand bluestem and blowout grass. (From *Grasslands of the Great Plains.*)

elsewhere on lower slopes intermixed with other grasses. In comparison with other grasses this one is shallowly rooted but the lateral spread of the roots is great. This mid grass renews growth early; it is very palatable to stock and produces more forage than most

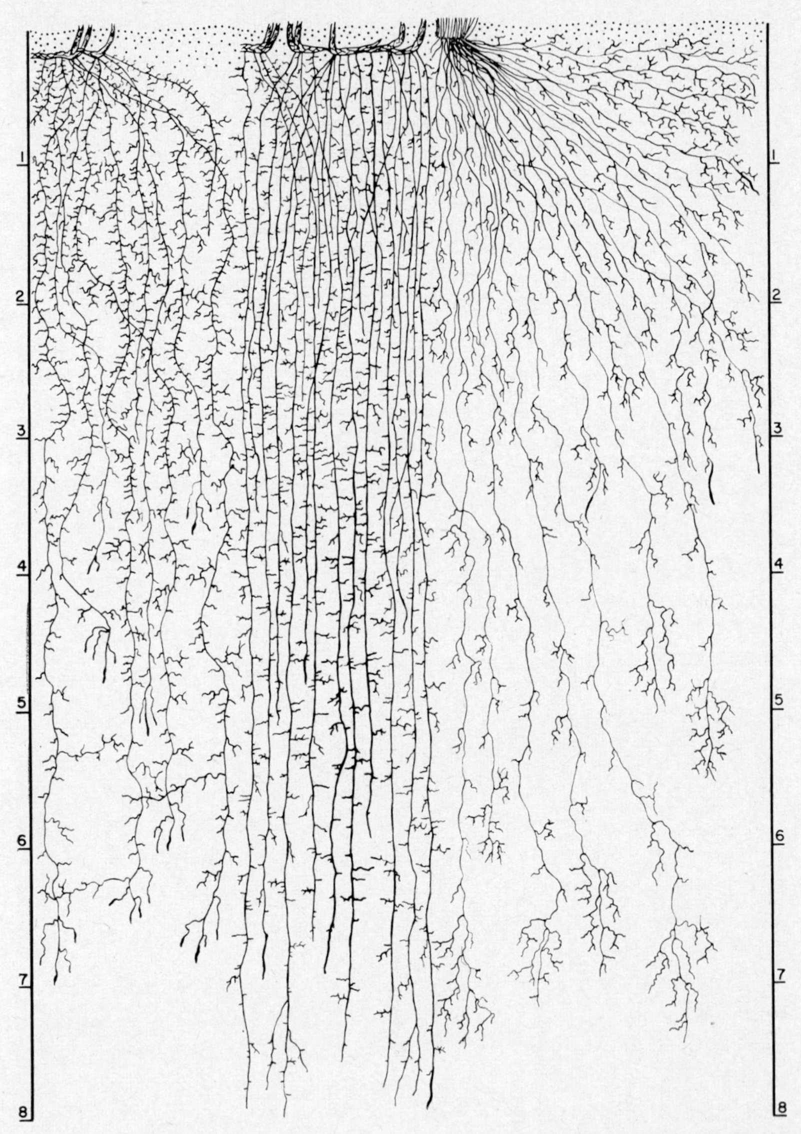

FIG. 122.—Roots of three dominant grasses of the sand hills. From left to right they are sand bluestem, sand reed, and little bluestem. Only the parts in a one-inch strip of soil are shown.

FIG. 123.—A clump of small soapweed which has held the sandy soil against severe wind erosion.

native grasses. Moreover, like buffalo grass and blue grama, it retains about half of its nutrients when winter cured.

Needle-and-thread about 65 years ago was ranked second in importance to little bluestem. Its tufts or small mats occurred thickly between the bunches of little bluestem and other grasses. But it is so palatable that, like needlegrass in true prairie, it has been grazed out in many areas. Needle-and-thread is especially abundant in semisandy soil in transition to hardlands (Fig. 110).

Junegrass, sand dropseed, black grama, and several sedges make up a large part of the understory. Indeed, short grasses such as blue grama and various sedges compose much of the ground layer. There are many contributing species but always a sparse cover.

There are several minor sand-hill communities, especially in the areas of transition from sandy to hardland soils. Needle-and-thread and Junegrass have been dominant over considerable areas north and west of the sand hills. Another was the wire-grass community where various three-awn grasses were dominant. It was best developed about the margins of the Sand-Hill Region on soils less sandy than that of the bunch grass community but not so firm as that of the hardlands.

Forbs

Although forbs are abundant and widespread, they are of secondary importance. Small soapweed is a characteristic species of mixed prairie (Figs. 123, 124). The roots of a representative group of forbs will be shown. It has been explained that this imaginary excavation 20 feet deep and 28 feet wide and long is merely a convenience for showing a number of root systems on one page. Indeed, the caving sand about individual plants created a considerable risk in excavating roots, but those of several plants of each species have been examined (Figs. 125, 126).

The much-branched and deep root system of lance-leaved psoralea (1) (Fig. 127) has been mentioned in considering blowouts. One of the spreading rhizomes is shown. This legume is very widespread. Conversely, the shallow but widely spreading roots of spiderwort (7) are also shown. Sand milkweed (9) is a common component of the vegetation but seldom occurs in abundance. Mature plants are 2 feet tall. The taproots are fleshy but branches are not abundant. The roots are sometimes 18 or more feet in depth. Bractless Mentzelia (3) is a tall, coarse, somewhat woody herb of wide distribution. It is often abundant locally and very conspicuous when the large yellowish-white flowers are in bloom during July and August. The poorly branched taproot is about 5 feet deep.

Sand sage (5) is a much-branched, bushy shrub a foot or two in width and 1.5 to 2 or more feet tall and of a dark green color (Fig. 126). It is especially abundant in the western and southwestern

FIG. 124.—Large hay meadow along a stream near Merriman.

FIG. 125.—Bush morning-glory with purple flowers. (From *Grasslands of the Great Plains*.)

FIG. 126.—Sand sage with needle-and-thread in overgrazed sand-hill range. (From *Grasslands of the Great Plains*.)

FIG. 127.—Excavation 20 feet deep showing root habits of several forbs of mixed-prairie sandhills. 1, lance-leaved psoralea; 2, slenderbush eriogonum; 3, bractless mentzelia; 4, whiteflower gilia; 5, sand sage; 6, hairy prairie clover; 7, prairie spiderwort; 8, bush morning-glory; 9, sand milkweed; 10, an evening primrose.

parts of the sand hills. It is found with sand reed and sand bluestem. Often over extensive areas in the sand hills it more or less replaces needle-and-thread. Under overuse of the range it increases in density of stand. Several ten-year-old plants were examined. They spread their roots 3 to 5 feet from the 1.5 inch woody taproot, especially in the upper two feet of soil. Branches were also abundant deeper and absorption occurs to a depth of 8 feet. This species is indicative of a light type of soil with considerable water penetration.

Slenderbush eriogonum (2) controls local areas on sandy slopes and occurs widely. This long-lived perennial is 1.5 to 2 feet high. Its root system is spread about 2 feet on all sides of the taproot but mostly in the first 3 feet of soil. The branches of the taproot reached even greater depths than did those of the sage. They were traced downward 10 feet. Repeated caving of the sand made digging so dangerous that the trench was abandoned.

The best developed root system of all was that of the bush morning-glory (8). This plant is widely distributed over the Great Plains and is a definite indicator of sandy soil (Fig. 125). Top

Fig. 128 (upper).—Sand cherry in fruit. This shrub is often common on the hills.

Fig. 129 (lower).—A volunteer crop of prairie sunflower on waste land.

growth consists of a bush-like plant distributed over several square feet. In summer it is covered with an abundance of large purple blossoms. A very large taproot, from 6 to 24 inches thick, extends downward, often tapering gradually to an inch or two in width at a depth of 4 to 5 feet. Here it may break up into very numerous fine roots. These abundant finer roots have been traced to a depth of 10 feet and some were probably much deeper. Lateral spread of many roots was at least 15 to 25 feet in all directions from the base of the plant. The root system is indeed magnificent. The enlarged portion of the taproot furnishes not only an enormous reservoir of food but it is also a storehouse of water. Average depth of this group of ten species was 6.5 feet.

Of the 45 species that were examined in sand hills only four may be designated as shallow-rooted, but each of these had a widely spreading root system. Among 23 species with roots of intermediate depth (2–5 feet) only four had roots which did not spread widely in the surface soil. The widely spreading, superficial root habit is a pronounced group characteristic. Of the 18 species of deeply-rooted plants, three of which are grasses, all but four had widely spreading laterals in addition to the extensive deep-seated ones.

In a comparison of root systems of mixed-prairie species of sand hills and hardlands, the only marked difference as a group is the somewhat better development of the root system in the sand, especially as regards smaller absorbing rootlets (Figs. 128, 129).

XIII.

Evergreen Forests and the Northwest

These are relatively small areas compared with the preceding topographic units. The vegetation, especially the forest, is distinctly different. It will be described only briefly.

Pine Ridge Region

In northwestern Nebraska, north of the Niobrara River, the Pine Ridge Region extends from Wyoming across northwest Nebraska and into South Dakota. It lies southeast of the Pierre Hills and gumbo plains and extends eastward to the Sand-Hill Region of Nebraska (Fig. 7). Geographically the Pine Ridge Escarpment separates the High Plains, which extend southward almost to the Rio Grande, from the Missouri Plateau, which stretches far northward into Canada. Its area is about 2,700 square miles. The topography varies from nearly flat to rolling, hilly, or almost mountainous. Much of the area is covered with a dense to medium stand of pine timber, and scattered trees are common. Pine Ridge is an irregular region of rough broken land, narrow valleys, and park land. It is a scenic region because of its high rugged topography and mantle of pines. The altitude is about 4,000 feet.

The average annual rainfall of about 18 inches, most of which occurs in spring, is sufficient for western yellow or ponderosa pine to grow on steep hillsides and in ravines (Figs. 130, 131). Summer drought is frequent, but the roots of the pine penetrate deeply and spread widely in the sandy, rocky soils. The frost-free period is short, only about 147 days, but the elevations are from 3,250 to 4,500 feet and thus sufficient to ameliorate the summer temperatures. Deciduous trees grow in canyons along the numerous small streams where their roots extend to the permanent supply of water through subirrigation during most of the summer. Here they are well protected from the direct force of drying winds.

Western yellow pine is the dominating species of the Pine Ridge Area (Figs. 132, 133). In 1915 it was stated that in Dawes County approximately 200 square miles were covered with a dense to

Fig. 130 (upper).—View of the Pine Ridge Escarpment. Yellow pine grows on the steep, rocky slopes; deciduous trees along streams; and mixed prairie covers the gentle slopes.

Fig. 131 (lower).—View on the Pine Ridge Escarpment west of Chadron. The trees are yellow pine alternating with mixed prairie. Photos Dr. W. L. Tolstead. (Fig. 131 from *Grasslands of the Great Plains*.)

medium stand of timber and that along Pine Ridge there was a moderately wide zone of scattered to dense pine timber. This extends along the rough land of both the north and south escarpments and the canyons crossing the park land between them. The trees are of medium size and rarely attain a height of more than 60 feet. The trunks are stocky with a short clear length and rarely exceed

FIG. 132.—A thick stand of ponderosa pine about 50 years old.

FIG. 133.—Another view in the pine forest with an old tree which escaped cutting.

a breast-high diameter of 2 feet. Very frequently in the open the branches of the pines extend almost to the ground. In areas of sparse timber growth there is considerable grazing land available. In some of the deeper canyons and on the north and east-facing slopes, a much denser forest aspect prevails. Here nearly all grass is excluded. The Pine Ridge country once yielded trees as large as 4 feet in diameter. The wood is used for lumber, posts, and fuel. As early as 1885, before the railroad extended into Dawes County, a sawmill was set in operation. Others followed later.

Small isolated stands of quaking aspen are occasionally found in the deeper canyons of Pine Ridge. They range from 15 to 30 feet in height and are 2 to 6 inches in diameter. These trees, with smooth bark and ashy-gray color, form a striking contrast to the coniferous species surrounding them.

In the northeast corner of Dawes County, the rough, stony topography, together with the abundance of lime derived from the underlying chalk rock, forms an ideal soil for Rocky Mountain or Western red cedar. The small amount of clay not completely removed, however, is detrimental to the western yellow pine which requires an open, sandy, well aerated soil. This red cedar is a short, gnarled, stocky tree seldom attaining a height of more than 20 feet, although the diameter a foot above the soil is 10 to 14 inches. The trunk has almost no clear length and the lower branches sometimes lie on the surface of the ground.

Pine Ridge in Dawes County has two well defined, north-facing escarpments separated by park land one-half to two miles wide. The upper escarpment forms the boundary line between Dawes Table on the south and Pine Ridge Park, which lies 150 to 200 feet below. The lower escarpment is along or near the line between Pine Ridge and the White River Basin. Pine Ridge Park has a gently rolling topography cut by deep narrow canyons. The divides are generally devoid of timber but the canyons contain fairly dense growths of pines. These trees form an arborescent network over the entire region. The treeless areas are never large, and seldom exceed 160 acres. They are so numerous, however, as to make this region more suitable for agriculture than for timber production.

Much of the forest in Pine Ridge was cut between 1880 and 1900. Recent studies (1947) show that the average age of the trees in many stands is now approximately 50 years; some are only 20 to 30 years old. Each stand is relatively uniform in age because most of the reproduction occurred within a few seasons after the cutting of the old forest. A few stands which escaped cutting are now 100 to 125 years old. Trees on the steep hillsides, ridges, and cliffs never have been used for lumber, because most of them are inaccessible and also because many have decayed, hollow centers. A few trees were found to be 150 to 200 years old. Only a few areas are covered with mature forest with 150 trees per acre, 60 to 90 feet tall, 14 to

FIG. 134.—Ponderosa pine in the canyon and an excellent development of needle-and-thread in the foreground.

18 inches in diameter at breast height, and 125 to 150 years old.

In dense stands of pines there is little undergrowth. But in open stands and between widely spaced trees, grasses and sedges and certain forbs of mixed prairie prevail. Thus in areas of sparse timber growth there is considerable grazing land. The most abundant grasses and sedges are blue grama, thread-leaf sedge, sun sedge, Kentucky bluegrass, needle-and-thread, little bluestem, and western wheatgrass (Figs. 134, 135). The most abundant shrubs are wolfberry, squawbush, wild rose, and poison ivy, but there are numerous others.

The rough land of the Pine Ridge area is used mostly for grazing and growing timber. The grasses and forbs found in this general area are the same as those of the surrounding hardlands and sand hills.

Species of deciduous trees are common in the draws and canyons of the Pine Ridge district. There is no real mixture of the two types of forest. A few pines and broadleaf trees grow together in the upper parts of water courses, but taken as a whole neither type encroaches upon the domain of the other. Deciduous trees grow in the subirrigated lands along creeks and drainages. Deciduous woods seldom occur in areas of more than 20 acres. The dominant trees are green ash, American elm, and boxelder. Hackberry is of minor importance. In well developed woods the trees are 40 to 50 feet high and about 9 inches in breast-high diameter. Quaking aspen in the deep canyons and mountain birch on banks of some streams are rare. Peach-leaved willow and cottonwood occur on the banks of small streams.

The canyon and valley floors leading from Pine Ridge across the slopes of the White River Basin are characterized by a fairly dense growth of broad-leaf species. The bottomland along the river is well timbered but the region north of the White River is timbered only very sparsely.

PIERRE HILLS

This area of about 600 square miles lies along the South Dakota line in Sioux and Dawes Counties (Fig. 7). It is part of a similar region extending into adjacent areas of South Dakota and Wyoming. The area is a rolling plain developed on soft clayey shales. The low hills are round-topped and the upland valleys are broad swales. A few low butte-like hills capped by gravel are present. Small areas with badlands topography are present north of Harrison, south of Orella at Toadstool Park, and south of the White River near Chadron. The area is covered with a thin but continuous stand of native grasses, principally western wheatgrass, needle-and-thread, blue grama, hairy grama, and buffalo grass. Cactus is often

FIG. 135.—A good stand of buffalo grass on Pine Ridge.

Fig. 136.—Portion of a thick stand of cactus on the Pierre Hills.

abundant (Fig. 136). There are no trees on the rolling hills. Willow, cottonwood, green ash, and elm grow along the perennial streams.

The land is used primarily for grazing by cattle. Well water for domestic purposes and use by livestock is difficult to obtain. Rain water from the roofs of farm buildings is caught and stored in cisterns for domestic use. Water for livestock is provided by small dams constructed across drainageways.

Wildcat Ridge

This area extends south-eastward about 70 miles from the Wyoming line. Its crest is 400 to 700 feet above the bordering valley of the North Platte River on the north and Pumpkin Creek on the south (Fig. 7). It is a region of bold topography; some of the buttes reach an altitude of 4,000 to 4,662 feet. There is an open forest of pine on the higher parts of the ridge while several species of deciduous trees occur on the canyon floor (Fig. 137). Thus the pine type of forest occupies two quite widely separated districts in Nebraska, where they have no connection, though there is a juncture farther west in Wyoming. This pine region has the same general characteristics as that of Pine Ridge.

The ponderosa pines extended eastward along the North Platte

River and Lodge Pole Creek to Deuel County, until the pioneers destroyed them more than 50 years ago. In places destruction has been so complete that scarcely a tree now remains.

The scarcity of timber in the surrounding country made severe demands upon the pine forest. Not far from Harrisburg, in Banner County, several sawmills were formerly operated, and large quantities of timber were cut in the later eighties and early nineties. A few sawmills were also worked in the canyons south of Gering, in Scotts Bluff County. The lumber produced by these mills was used by the settlers in the construction of their houses and for many local purposes. Of the red cedar found with the pine, a large amount was used for fence posts, but that too is nearly all gone.

Much of western Nebraska is really a foothill region, and in it eastern and western species mingle. Western yellow pine occurs on the bluffs of the Niobrara River eastward to Keya Paha and Holt Counties as well as in the rough country of Scotts Bluff, Banner, and Kimball Counties, and in the breaks of the North Platte River eastward to Morrill and Garden Counties. Isolated groves also occur eastward in the valleys of several rivers. Limber pine occurs in small groves in Kimball County. Several shrubs are also of western origin.

FIG. 137.—A view of the Wildcat Ridge with ponderosa pines on the rough land.

XIV.

Cultivated Crops of Grassland Soils

Wheat, rye, sorghums and other cereal crops are grown widely throughout the plains and tall-grass prairie. Field corn is a chief crop where true-prairie land has been broken. We will now examine the growth of these crops in relation to soils and climate and then compare their relative demands for water with that of the native vegetation.

Growth of Cereals in Mixed and True Prairie

Growth of winter wheat and winter rye was examined in 1919 at 13 different stations. These extended throughout the mixed and true prairie from near the Rocky Mountains nearly to the Missouri River. Six were in mixed prairie and seven in true prairie. This study was made for the Carnegie Institution of Washington. Although the soils, preceding crops, etc., were described at each station, it is only necessary here to state that the crops were grown by farmers under the usual methods of soil preparation and seeding common to the district in which each crop was produced. A statement of grain yield would also have been very desirable had it not involved either an amount of work incommensurate with its value, or data furnished by a number of farmers and ranchmen. Moreover, as is well known, local conditions at the time of blossoming or when the grain is maturing often materially affect, if indeed they do not largely determine, the yield. Undoubtedly the vegetative development was the best criterion for our purpose. All crops were grown on non-sandy land (Figs. 138, 139).

Development of the cereal crops in feet at the widely separated stations follows.

The relative development of the cereals at each station showed a striking relationship between the growth of the crop and the degree of xerophytism of the plant community. Average height of winter wheat was 1.3 feet greater in true prairie, where mid and

Station	Height tops	Depth roots	Height tops	Depth roots
		Mixed Prairie		
	Wheat		*Rye*	
Limon, Colo.* 13.4	1.8	2.8	2.3	2.0
Flagler, Colo.	1.0	1.5	2.1	2.8
Sterling, Colo.* 18.1	2.0	2.8	...	...
Yuma, Colo. (20.2)	2.1	2.3	2.8	2.8
Ardmore, S. Dak. (16.7)	2.6	4.1	...	...
Colby, Kans. (22.5)	3.2	2.3	3.5	3.6
Average	2.1	2.6	2.7	2.8
		True Prairie		
Mankato, Kans.	...	...	4.2	4.7
Lincoln, Nebr. (29.3)	3.3	4.7	5.5	5.0
Lincoln, Nebr. (29.3)	3.8	6.2	6.5	5.0
Lincoln, Nebr. (29.3)	3.5	7.3	...	...
Wahoo, Nebr.	3.5	5.0	3.5	5.0
Fairbury, Nebr. (33.4)	3.0	4.1	4.5	5.2
Fairbury, Nebr. (33.4)	...	...	3.9	4.2
Average	3.4	5.5	4.7	4.9

NOTE: Numbers in parentheses following names of stations indicate precipitation (where available) from July 1918 to July 1919. An asterisk denotes mean annual rainfall. Stations near Lincoln and Fairbury were several miles apart.

FIG. 138.—Field of wheat at Colby, Kansas, 1919, where roots were excavated.

FIG. 139.—Wheat field at Lincoln in 1919.

tall grasses prevail, than in the mid and short grasses of the Great Plains. Average depth of roots was also 2.9 feet greater. Similar differences for winter rye were 2 feet for tops and 2.1 feet for root depth.

At Burlington, Colorado, due to excessive previous rainfall, the soil was quite moist to a depth of 7.5 feet and was easily molded into firm lumps with only slight pressure of the hand. Here mean annual rainfall was only 17.2 inches. Wheat was 3.5 feet tall and maximum root depth was 6 feet; both figures were quite above the average. Similar occasional findings of deep moist soil in various other places in the Great Plains offer an explanation for the deep root penetration of native plants. At some time or times during their long lifetime they have been able to extend their roots deeply, and to maintain this advantage. Short-lived cereal crops (for corresponding results were had for oats) made their least vegetative development throughout the mixed prairie and their best development in true prairie. As is well known, height of cereals has little effect on yield.

The great amount of runoff and the high evaporation rates are factors to be kept constantly in mind in evaluating efficient rainfall. Water-content of soil, and not soil fertility, is the chief limiting factor in mixed prairie. These data emphasize the depth at which crop plants distribute their roots when grown under favorable climatic conditions, as indicated by true prairie, where the subsoil is moist. Such studies must lead to a revision of the current ideas (common in the first quarter of the century) of the depths at which crop plants carry on absorption, especially when nearing maturity. The smaller development of both aerial and underground portions of crops in mixed prairie than in true prairie was to be expected,

FIG. 140.—Plants of spring-sown barley from Lincoln, Nebraska (left); Phillipsburg, Kansas; and Burlington, Colorado (right) on June 10, 1921. Seed was planted on March 24 to 30.

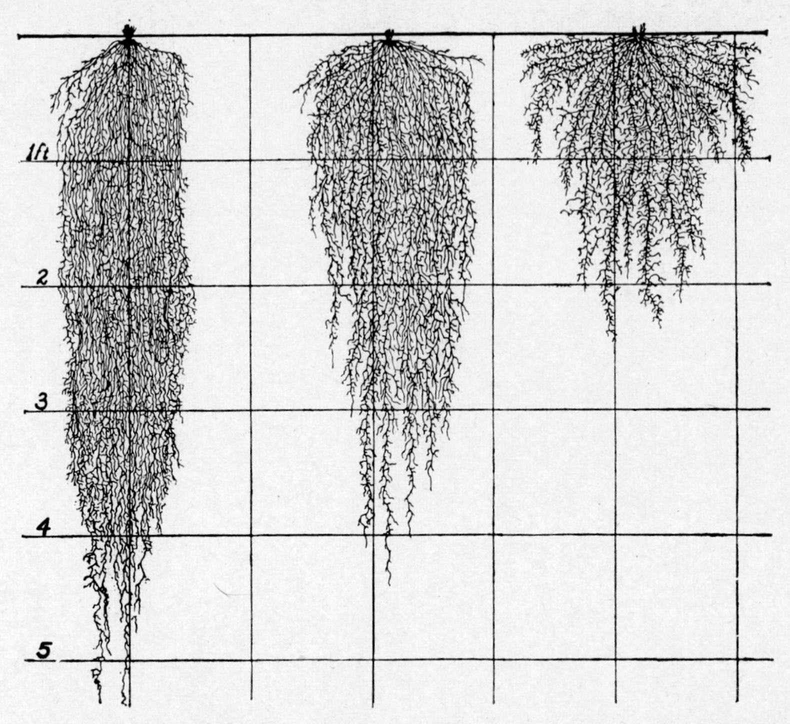

Fig. 141.—Average root development of winter wheat in fertile silt loam soils under a precipitation of 26 to 32 inches (left), 21 to 24 inches (center), and 16 to 19 inches (right). Note the wider spread of roots in drier soil. Average from studies made in 20 fields.

but the correlation had not heretofore been worked out (Figs. 140, 141).

Development of Root and Shoot of Winter Wheat

During the fall of 1921 a study was made of the development of a strain of Turkey Red winter wheat (Kanred) both above and below ground. The crop was grown in a deep, fertile, silt loam in a field near Lincoln. The roots of the preceding crop of corn, which had been harvested in August, decayed rapidly, and all stubble and debris were removed to prevent their interfering with root excavations. Wheat was drilled on September 20 at a depth of 1.5 to 2 inches, and at the usual rate of 75 pounds per acre. Favorable environmental conditions promoted prompt germination and growth. The soil was continuously moist at all depths. Roots were excavated at 10-day intervals, from September 30 to November 29, and a typical root system was drawn to scale.

The roots consisted of a primary system of three roots which penetrated rather vertically downward, at the rate of more than half an inch per day, and, branching widely, reached depths of 3 to 4 feet by mid-December. Production of tillers began 15 days after planting. A new tiller was produced every four or five days until December, at which time an average of 15 tillers and 43 leaves per plant was attained. Even in November, growth of roots progressed rapidly at average daily air temperatures of 34° F.

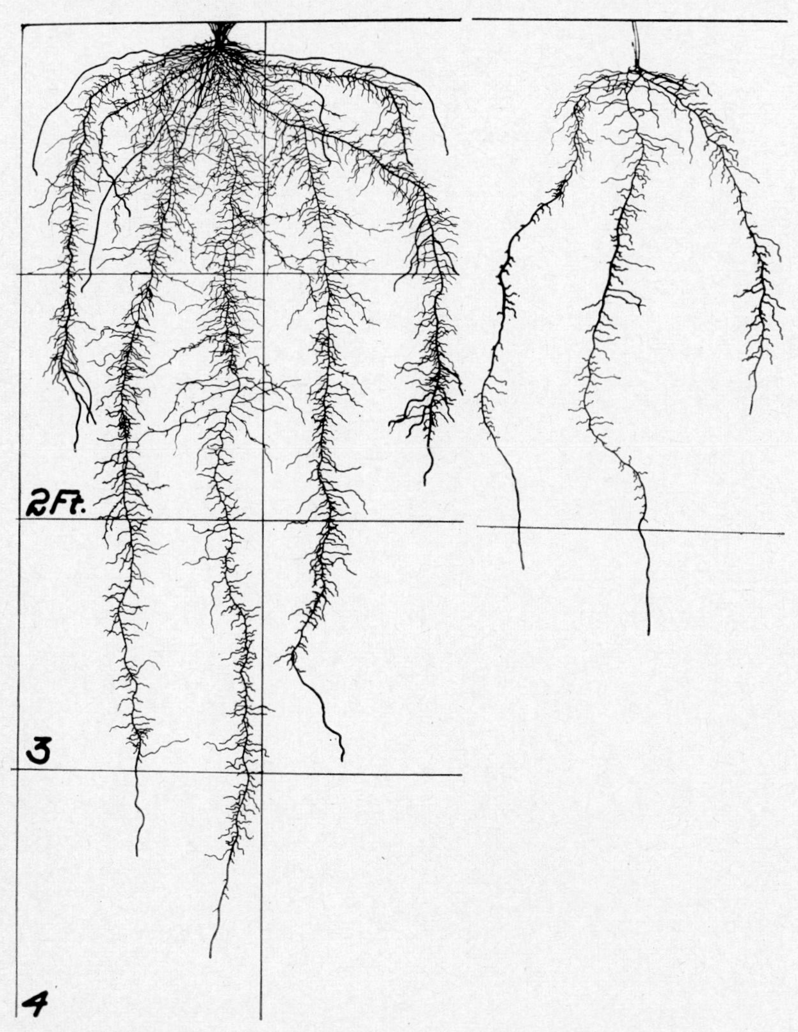

FIG. 142.–(Right) Root of Kanred winter wheat grown at Lincoln when 20 days old; (left) root of similar plant on November 29, when 70 days old.

Development of the secondary root system, as in native grasses, began at the same time as the appearance of tillers (Fig. 142). It was correlated with that of tiller production; a new root was added every four or five days until mid-November. Thereafter chief root development was not in number but in elongation and branching. By December, the secondary root system had an average of 11 roots, some of which were 22 inches deep. The primary root system had spread about a foot on all sides of the plant and, turning downward, branched so freely in the upper soil that it furnished half of the total absorbing system exclusive of root hairs.

Rate of Growth of Field Corn

Half of a plant and often much more is frequently hidden from view; roots often extend quite through the soil or solum and deep into the parent materials. In the great majority of both native and cultivated plants the root systems are, in comparison to tops, deeply penetrating and very extensive. In annual plants this necessarily requires a rapid rate of growth. An excellent example is that of field corn (Fig. 143).

Fig. 143.—Deep trenches must be excavated in a study of field crops.

After plowing and harrowing a field in 1919 at Peru, Nebraska, Iowa Silver Mine corn, a late maturing variety, was drilled in rows 3 feet apart with kernels 17 inches distant in the row on May 9. The rich loess soil was deep and moist throughout. After 36 days the plants were 10 to 14 inches tall and the seventh and eighth leaves were unfolding. All of the wonderfully branched roots of both the primary and secondary root systems occurred in the surface foot of soil and extended laterally about 2 feet on all sides of the plant.

A second examination on July 5, when the plants were 57 days old and about 4 feet tall, revealed that a remarkable extention of the root system had

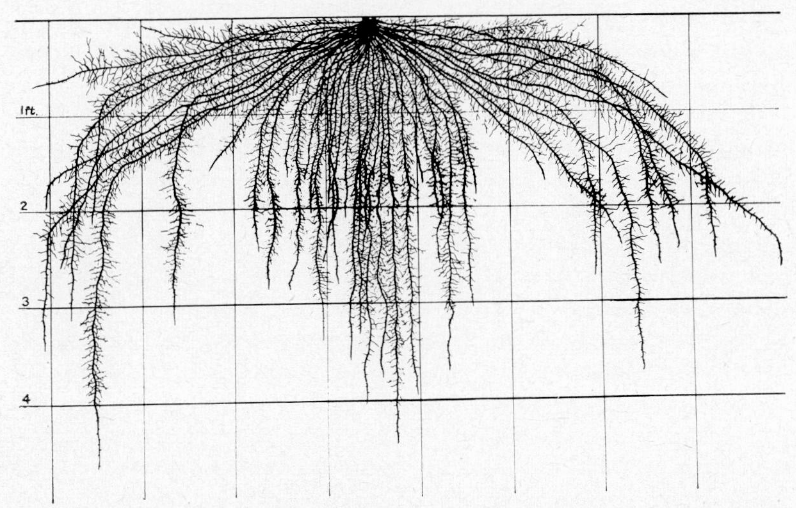

FIG. 144.—Roots of Iowa Silver Mine corn grown in loess soil at Peru, Nebraska, on July 5, eight weeks after planting.

occurred. In Figure 144, which shows all of the roots projected in one plane, it may be seen that one group of roots extended outward 2 to 3 feet on all sides of the plant and only then turned downward. A second group, the youngest roots, spread less widely but pursued a somewhat vertically downward course. Note the effect that cultivation to a depth 3 to 4 inches, a type that was general in those days, would have had on this remarkable expanse of roots. In fact, following this discovery of root position, the J. I. Case and other implement companies replaced the few large shovels on their cultivators with numerous smaller ones which moved only the very top soil.

The corn was growing rapidly; it had a great leaf surface; and demands for more water and nutrients were increasing daily. During the 21 days since the first examination, average growth in height was 1.7 inches per day.

The roots of maturing corn plants were examined on September 2, nearly four months after planting. The stalks were 8 to 9 feet high and, although a few leaves had dried, most of them were still green. The husks on the ears were just beginning to dry and the kernels were dented. Plants had completed their root growth. The shallow portion of the root system had scarcely increased over that found at the previous examination, so far as lateral spread was concerned. Some roots had extended a little deeper; others maintained a horizontal position. The roots were profusely branched to the second and third order with branches ranging from less than an inch to 2 feet in length. The surface soil to a depth 8 to 12 inches was filled with these fine rootlets (Fig. 145).

Unlike the shallower portion of the root system, the more deeply penetrating part had made a remarkable development. The number of the more vertically penetrating roots varied from about 20 to 35 per plant. Of these a few were short and did not reach a depth greater than 1 to 2 feet; many were 7 feet deep, while the longest penetrated to 8.2 feet. All were profusely branched with laterals varying in length from less than an inch to 15 inches. Often 10 to 12 branches were found on an inch of the main root. To attain such depths the main vertical roots penetrate downward, under very favorable conditions, at the remarkable rate of 2 to 2.5 inches per day during a period of 3 to 4 weeks. Thus, field corn is furnished with a remarkably extensive and efficient root system. The plants "stick" down as far as they "stick" up.

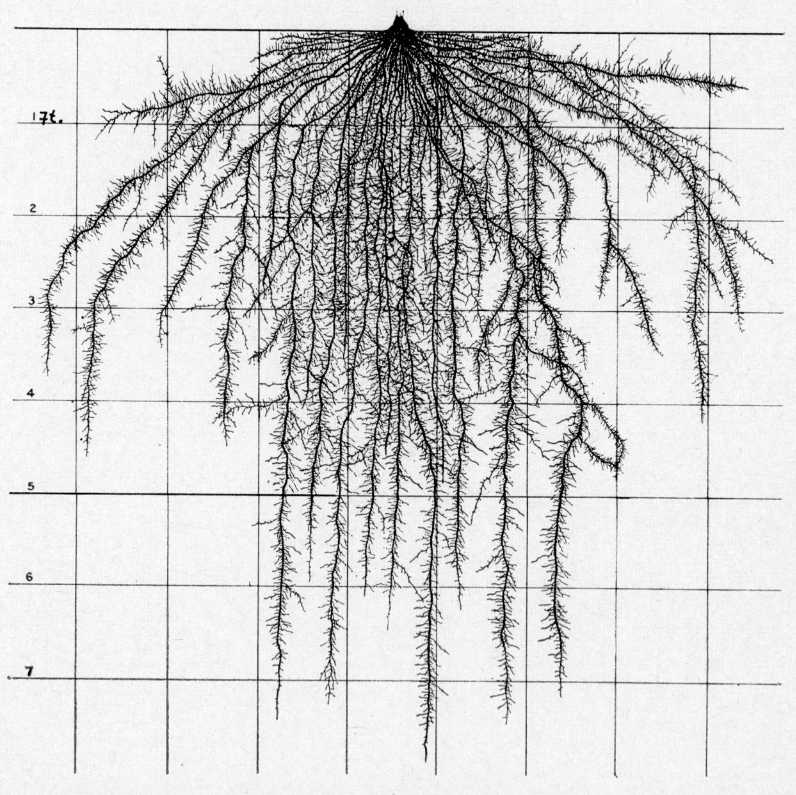

FIG. 145.—Mature root system of corn in September.

Sorghums

All of the many varieties of sorghum are annuals. They have a well-developed root system similar to that of corn but generally finer and more fibrous. Sudan grass, grown at Lincoln, reached a height of 7.5 feet, and many roots penetrated to a depth of 6.8 feet (Fig. 146). Extensive studies on various sorghums, kafir corn, and milo grown in the same field and often in alternate rows with corn have shown a great similarity in number and extent of roots to those of corn. But the number of secondary roots was approximately twice as great as the number on corn. Moreover, the sorghums had leaf areas only about half as great as those of corn. Thus, the excellent root system coupled with a relatively small transpiring area and a low water requirement, goes far towards explaining the high degree of resistance of sorghums to drought. Their ability to remain in an almost quiescent state during drought is another important characteristic. The sorghums remain fresh and green during periods of dry weather which would be extremely harmful to corn. In fact, the plants may even cease growth for a considerable time, but when revived by a rain a vigorous growth rate is resumed.

Depths at Which Plants Absorb Water and Nutrients

Studies on the rooting habits of crops, giving a clear understanding of their extent, were not made until about 1922. At that time the old idea of roots being superficial was commonly believed, and many textbooks spoke of them as shallow. This old viewpoint was well stated as follows in one of the best modern works on soils. "It is well known that only the top 6 or 8 inches of soil is suited to plant life, and that the lower part, or subsoil, plays only an indirect part in plant nutrition. We shall therefore, confine our attention almost exclusively to the surface layer."*

The great extent of the root systems of most plants that have been examined and their usual thorough occupancy of the subsoil have led to investigations concerning their activities in subsoil and parent materials. Although the capillary movement of water in soils had by this time (1922) been shown to be much less than formerly supposed, experiments were devised to check all movements of either water or nutrients from one soil layer into another, except through the roots. In filling the containers, the soil, after being brought to the desired water-content, was firmly compacted and

*E. J. Russell, *Soil Conditions and Plant Growth* (London: Longmans, Green and Company, 1917).

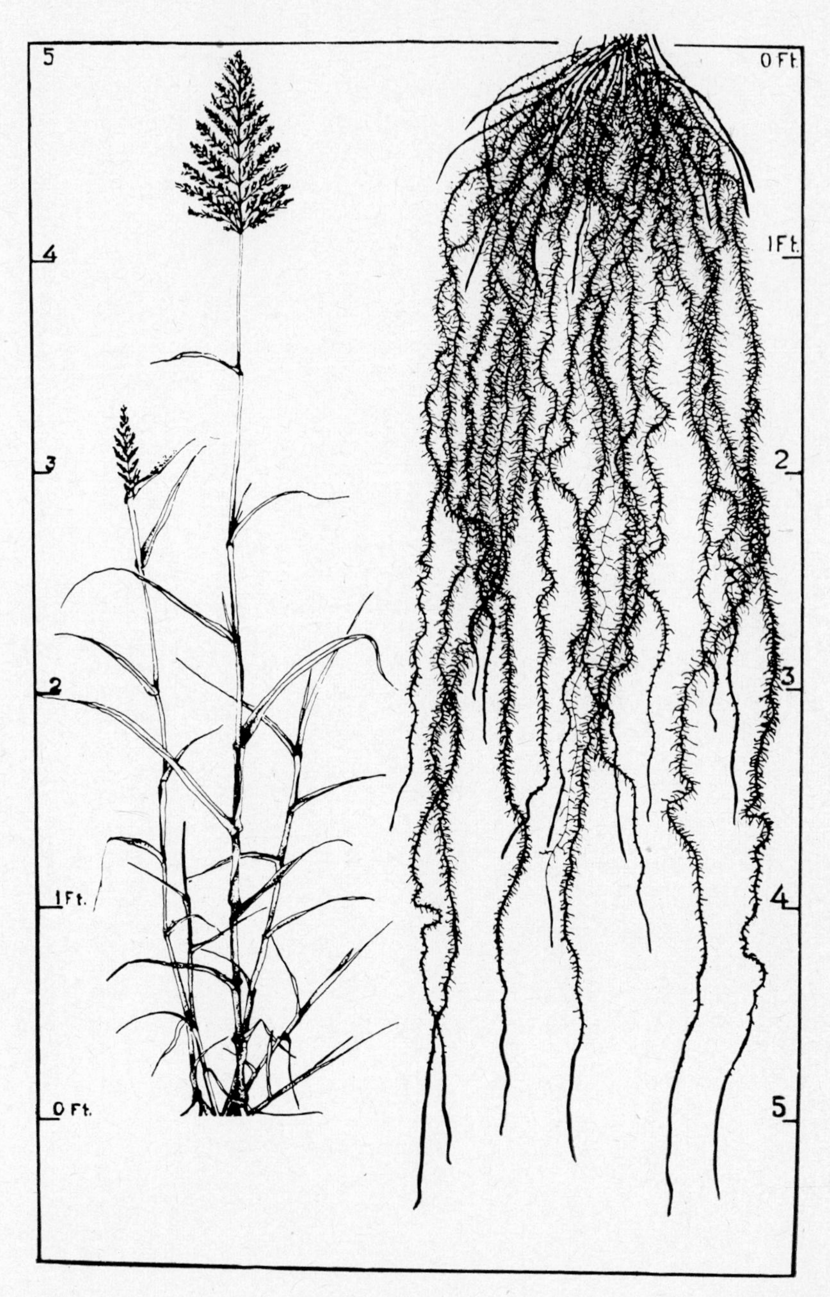

Fig. 146.—Tops and roots of Sudan grass grown at Lincoln.

then sealed with a layer of wax. This consisted of 85 percent paraffin or parowax and 15 percent petrolatum. It was applied hot so that it penetrated a little into the soil, and when it cooled clung tenaciously to the soil particles. The seal was only 2 to 3 mm. thick. When it had cooled and hardened, another layer of soil, usually 6 to 12 inches thick, was added and the process repeated until the container was filled. The roots of various native and crop plants grown in these soils were distributed evenly throughout the several hermetically sealed layers, penetrating the wax seals without difficulty. The seal had no apparent effect on root development.

This method furnishes at once a means of determining the amount of water or nutrients removed at any given level to which the roots penetrated. Moreover, by using a series of containers in which plants of the same age were grown, it was possible, by opening containers from time to time, to determine the absorbing activity of the roots at the various levels at any stage in the development of the crop. In these experiments, containers 1.5 to 3 feet in diameter and 2.5 to 5 feet deep were employed. They were placed in trenches which were then refilled with soil, and crops planted around the containers in such a manner that the experimental crops were grown under field conditions. Moreover, the containers were filled so that the well-compacted soil at any level occupied the same relative position that it had occupied before removal from the field. To prevent water intake, each container was furnished with an appropriate wooden roof. During a period of two years, 50 containers were used.

In one experiment with barley, for example, it was found, when the crop was ripe, that water had been absorbed from the several six-inch layers in the following amounts: 20, 19, 16, 14, 12 and 11 percent, respectively, based upon the dry weight of the soil. It was also ascertained, from barley grown in other containers and examined at various intervals, that during the heading and ripening of the grain, a most critical period in plant development, the bulk of absorption was carried on by the younger portions of the roots in the deeper soil.

Corn was found to be an extravagant user of water, absorbing large quantities from the third and fourth foot of soil and smaller amounts from the fifth foot. Potatoes absorbed water to depths of their root extent, about 2.5 feet, and western wheatgrass and big bluestem, grown from transplanted sods, showed marked absorption to a similar depth.

In other experiments measured amounts of nitrates (400 parts per million of soil) were placed in the soil at various depths. Barley, when 2.5 months old, had removed 286 and 135 parts per million of nitrates from the 1.5 and from the 1.5 to 2 foot soil levels. It also absorbed 168 ppm from the 2.5 foot level and, at maturity, 186 ppm from the 2.5 to 3 foot level. The two native grasses used nitrates from the second and third foot in large amounts, while corn removed 203, 140 and 118 ppm, respectively, from the third, fourth, and fifth foot of soil.

These experiments revealed for the first time the absorption of water and nitrogen at great depths. It was thus pointed out that the student of environment must consider not only the conditions in the surface soil but also the whole substratum to the depth of root penetration.*

A study of absorption of nutrients from subsoil in relation to crop yield followed in 1924.** The same methods as those described were used, and barley was grown in the field, in 18 large, water-tight, oak barrels. The soil, except in the controls, was fertilized at various levels to 2.5 feet in depth, and in some containers at two levels, with either a nitrate or phosphate fertilizer. A control container without a crop or fertilizer, but sealed at every 6-inch level, was used as a check to determine nitrification or denitrification.

Roots of the control plants often extended to the bottom of the containers, which were 2.5 feet deep, but nitrate fertilizer at any level lessened root depth and greatly increased branching. Both nutrients were absorbed in large quantities at all levels; although the greatest amount of salts were absorbed in the surface foot, the plants took additional quantities from the deeper levels when it was available.

Absorption of these nutrients below the upper foot materially affected both the quantity and quality of the yield. This occurred even when the surface foot was abundantly supplied with a similar nutrient. Total dry weight of the plant was increased when nitrates were applied in the upper soil early in its life. This increase resulted from heavier tillering. Nitrates increased dry weight and also quality of the grain still more when they were also available at levels below the surface foot. Conversely, phosphates depressed yield, especially that of the straw, somewhat in proportion to the amount absorbed.

The effects of nutritive salts are most marked not only early but

*These data are from *Development and Activities of Roots of Crop Plants*, Publication 316, Carnegie Institution of Washington (1922).

**These data are from the *Botanical Gazette* (April 1924).

also late in the development of the plant, when the longer roots are absorbing from the deeper soil levels. The ample distribution of the deeper portion of the root system in a rich subsoil solution at the latter critical period of growth is exceedingly important. The fact that roots absorb nutrients at deep levels in the subsoil as well as from the surface layer should be given greater attention by all plant growers.

Roots of Vegetable Crops

The underground parts of vegetable crops had received relatively little attention until 1927, and indeed accurate information was rarely to be found. The roots of plants were the least known, least understood, and least appreciated part of the plant. This was undoubtedly due to the fact that they are effectually hidden from sight. Notwithstanding the extreme difficulty and tediousness of laying the roots bare for study, it is not only remarkable but also extremely unfortunate that such investigations had been so long delayed. The student of plant production should have a vivid, mental picture of the plant as a whole. It is just as much of a biological unit as an animal. But with plants our mental conception is blurred by the fact that one of the most important structures is underground.

In 1926 a large experimental garden was established at Lincoln. Vegetable crops were grown in a deep, fertile, silt loam soil. The soil was not only in an excellent condition as regards tilth but it was also well supplied with humus. In spring it was moist at all depths examined, and at all times there was water available for growth. During the six months of the growing season there were 19.2 inches of rainfall.

Crops were grown in triplicate plots large enough to permit normal field development; each group of plants was entirely surrounded by plants of its kind. Cultivation was limited to a depth of an inch; weeds were kept out. Root development of each of the species was examined at three intervals during its development. Thirty-three species of vegetables were grown. Brief descriptions of two annual plants will be given.*

Tomato seedlings of the John Baer variety were transplanted to the Lincoln garden in mid-May. Omitting an earlier examination, by July 11 the plants had an average height of 19 inches, a spread of tops of 2 feet, and each was furnished with 6 to 11

*Data from Weaver and Bruner, *Root Development of Vegetable Crops* (New York: McGraw-Hill Book Company, 1927).

branches. Average total leaf surface was approximately 12 square feet. The plants were blooming profusely and had also developed a few small fruits. The taproot, as is usual, had been broken in transplanting but lateral branches from the six-inch stub were profuse. Many of these had extended outward 2 or more feet in the surface 9 inches of soil. Then, turning obliquely downward, they had reached depths of 3 to more than 3.5 feet. Other main roots with only a little lateral spread also extended downward. All roots were profusely branched (Fig. 147).

By mid-August the well-developed plants had a spread of tops of 4 feet. Plants with 294 leaves had a transpiring surface of 69 square feet. They were fruiting abundantly. Usually 15 to 20 major branch roots spread widely (maximum 5.5 feet), had very numerous large laterals, which usually penetrated deeply, and finally, turning downward, extended into the third, fourth, and sometimes the fifth foot of soil. In addition there were almost countless smaller main laterals. All were furnished with masses of much rebranched rootlets so that the absorbing system was extremely intricate and extensive to depths of 4 feet. Roots also arose from prostrate stems where they contacted the soil. Thus a soil volume with a surface area of at least 80 square feet and a depth of 3.5 feet—280 cubic feet of soil—was fully occupied by the roots of a single plant.

Members of the cucurbit, vine crop, or gourd family deserve special attention since their root systems are of an unusual pattern. They included, in this study, cucumber, muskmelon, watermelon,

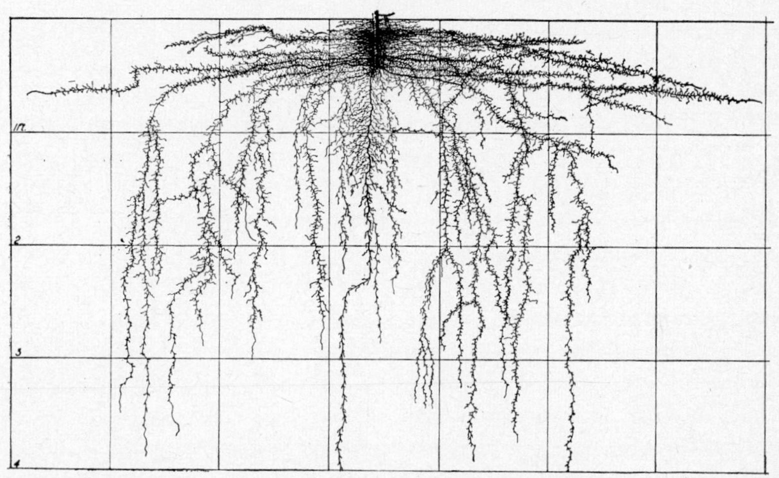

FIG. 147.—Roots of tomato (John Baer variety) on July 11, when about two-thirds grown. Roots of vegetable crops are from plants grown at Lincoln.

pumpkin, and squash. The root system of squash reveals the general pattern of the group.

The Golden Hubbard squash (*Cucurbita maxima*), like the other cultivated members of this family, is an annual. It has long, running, cylindrical stems which are often 12 to 24 feet in length and root freely at the nodes. When six weeks old and with a leaf surface of 3.4 square feet, a depth of 2.5 feet had been attained and numerous widely spreading laterals, mostly about a foot long, extended outward almost horizontally in the surface 18 inches of soil.

Two weeks later, when the plants had begun to bloom, a second examination was made. The main stem had two branches and 20 large leaves with blades about 9 by 8 inches in length and breadth, and 8 smaller ones half that size. The plants were making a vigorous growth. The general plan of the root habit had not changed but the root system was more extensive and elaborate. The taproot had increased only a foot in depth but the lateral spread of its branches had increased to a maximum of 5 feet. This spread was attained in the surface 8 inches of soil. In fact the chief growth was that of laterals which had extended more widely, were more profusely branched, and had many more longer branches than before. The degree of ultimate branching had also increased.

At the final examination on August 21 the plants were in the late blooming stage. Nine fruits 2 to 7 inches in diameter were found. The plant selected for study had two main vines. One of these was 18 feet long and had five branches which were 3 to 7 feet in length. The other one was 21 feet in length with 7 branches; the longest was 7 feet. These vines were furnished with two rows of leaves which averaged a foot in length and width. They thus presented an enormous food-making and transpiring surface, and probably lost many gallons of water per day.

The taproot and its deeper branches had made considerable growth and were extensively branched to the 6-foot level. The branches were, however, relatively short so that below 2 feet not a very large volume of soil was occupied.

The most important, most active, and by far the most extensive portion of the root system occupied the surface soil. Five to seven main shallow roots were common. These occurred usually at a depth of only 6 to 8 inches and were never found below the 12-inch soil level. They spread on all sides of the plant and branched profusely except in the first 12 to 18 inches of their course. They were still growing at the rate of nearly 2.5 inches per day (Figs. 148, 149). Please note that the scale is in square feet.

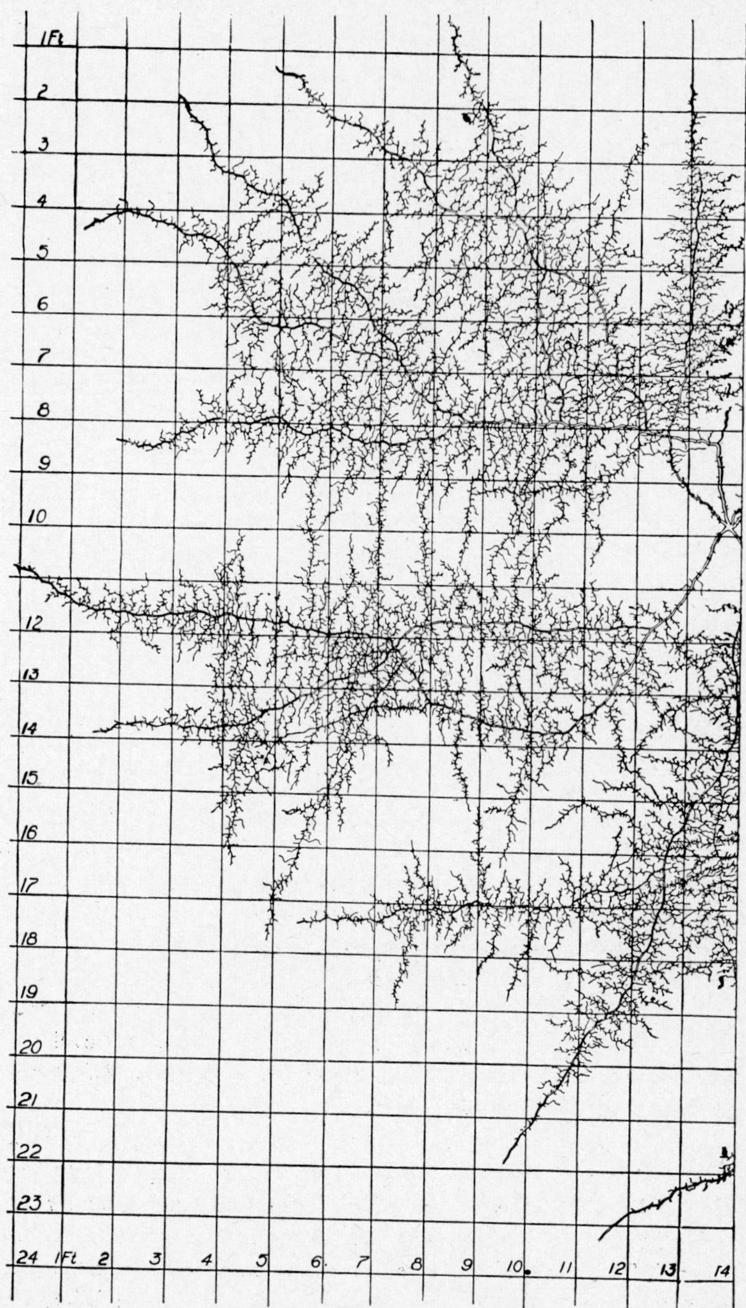

FIG. 148.—Surface view of one-half of the root system of Golden Hubbard squash on August 21. Many roots were growing at the rate of 2.5 inches per day.

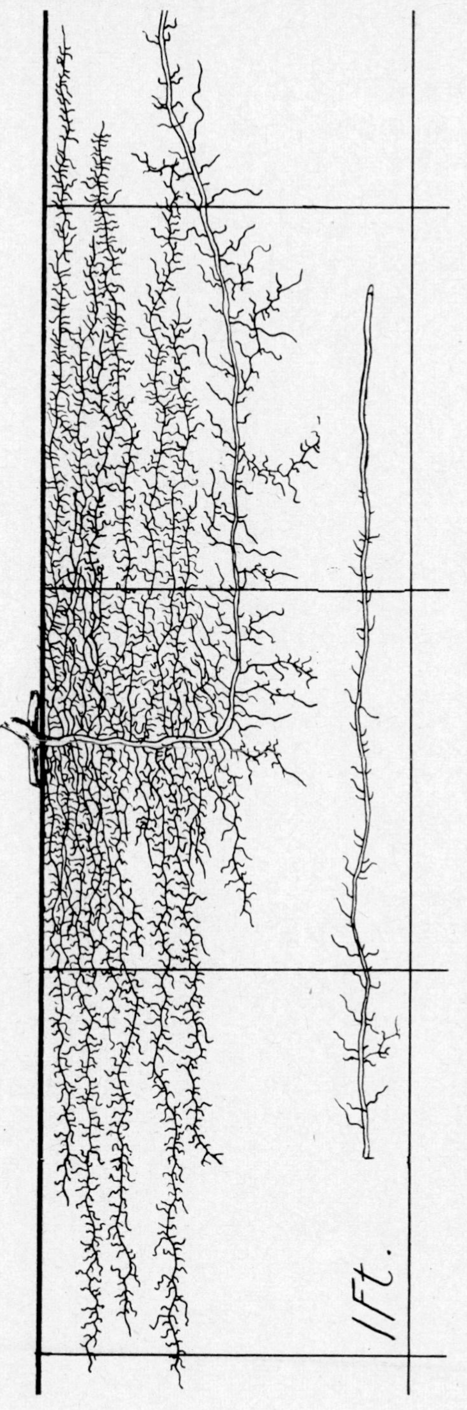

Fig. 149.—One of numerous roots originating from the nodes of a squash vine. These added greatly to the total absorbing area.

To ascertain the rate of growth, the soil 10 to 15 feet from the hill was carefully examined until the uninjured, glistening white, smooth ends of the main roots were found. Each was immediately engaged in a loop of stout cord and the root covered quickly with soil. The cord was tied to a small stake driven by the side of the root, the upper end of the stake being plainly visible and marking

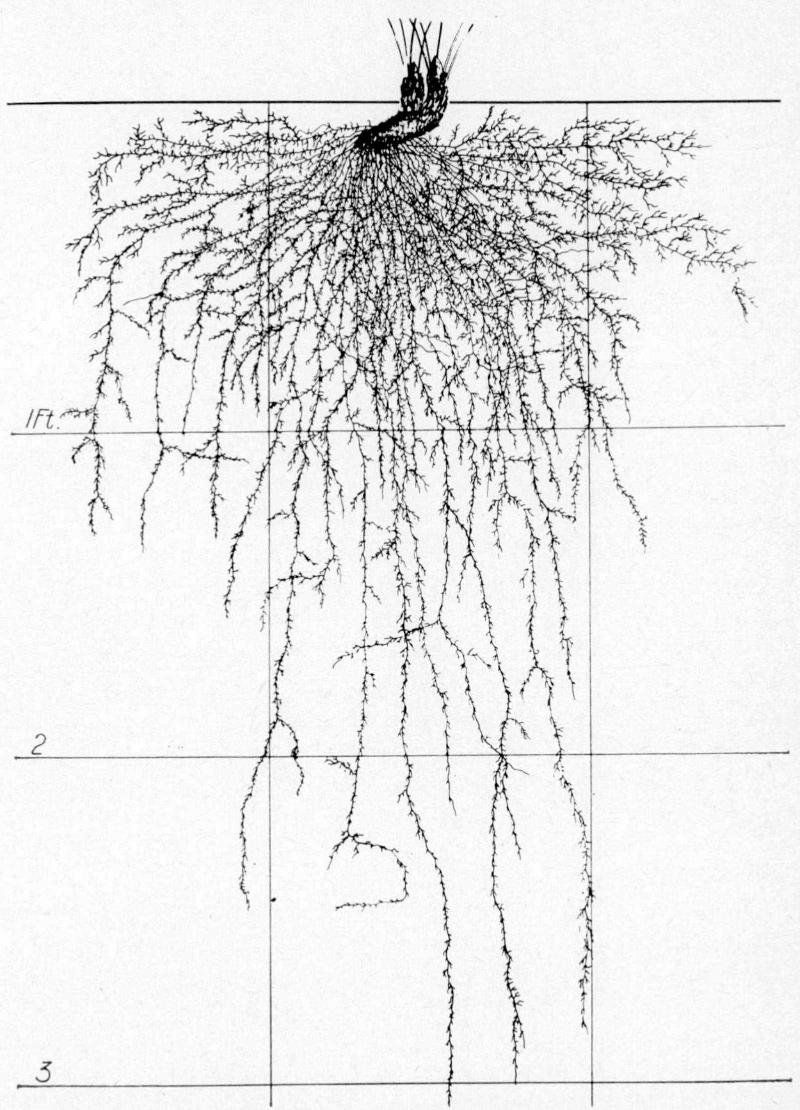

FIG. 150.—Underground parts of a 3-year-old Dunlap strawberry excavated in June after blossoming and fruiting.

the position of the root when the soil was replaced. The positions of several roots were thus marked on August 22. A week later an examination showed that some of the roots had made a growth of 17 inches in length, nearly 2.5 inches per day. Moreover, the unbranched root-ends were now clothed with branches some of which were 2 inches long.

It is evident that Figure 148 is incomplete. A mature root system filled the soil to a distance of 15 feet on all sides of the plant. In addition to the roots described, numerous nodal roots, also shallow and wide-spreading, were developing. Several hundred cubic feet of the richest soil may be occupied by the roots of a single plant of squash, pumpkin or watermelon.

Roots of a few perennial plants were examined. A four-year-old plant of rhubarb occupied more than 50 square feet of soil to a depth of 8 feet. Roots of a 10-year-old plant of horseradish filled the earth beneath 9 square feet of soil to a depth of 14 feet. It was abundantly supplied with profusely branched laterals. Conversely, strawberry of the Dunlap variety when three years old was the most shallowly rooted of all. The bulk of the roots occurred in the first foot of soil, where lateral spread was only about 1 foot. Very few roots reached depths of 3 feet (Fig. 150). This and the two preceding plants were examined at Lincoln but not in the experimental garden.

The following list, all annuals, is representative. The depth and lateral spread, in parentheses, are given in feet.

Onion	3 (1)	Radish	6 (3)
Garden beet	9 (4)	Pea	3 (2)
Cabbage	5 (3.5)	Lima bean	5 (4)
Turnip	4 (1.5)	Carrot	7.5 (2.5)

These are remarkable depths for a single season of growth and could be attained only in deep, rich soils, such as occur in Nebraska (Fig. 151).

Relative Demands for Water

The prairie vegetation, rooted throughout several feet of soil, begins growth early in spring. The field for corn, on the contrary, lies fallow. Planting time occurs in May, and the new crop may absorb only in the surface foot until well into June. The prairie continues growth notwithstanding drought in the surface soil, drought that may be so severe as to check greatly the growth of the new corn. With the progress of the season, the need for more and more water by the developing foliage in the prairie increases. This

FIG. 151.—The wonderful prairie sod is broken, but the rich soil supports other grass crops, Lincoln, 1932.

is in accord with the annual rainfall which usually reaches a maximum in June or July. These demands do not all occur at the same soil level. They are met in part by direct absorption from the soil even far below 5 feet. By midsummer a maximum transpiring area has been attained by the dominant grasses; the increase of water demands by the still growing autumnal forbs are probably largely offset by the waning or disappearance of vernal or estival species. Not so in the field of maize. The need for water constantly increases until flowering and fruiting. Roots are rapidly extended into the deeper soil—into the third, fourth, and fifth foot, which heretofore has furnished little or none of the supply. The demands for water are great and urgent. Actually these deeper soil layers are more depleted of their moisture early in autumn than are the corresponding ones in prairie.

Water relations in a field of winter wheat are quite different from those in either cornfield or prairie. When they begin their growth in the surface layer in early fall, the wheat roots often reach a depth of 3 to 3.5 feet before growth is retarded or ceases because of low temperatures. In spring, resumption of growth precedes that in the prairie, and the maximum demands for water—by all plants at the same levels—precedes that in the grassland. By midsummer it ceases abruptly.

The prairie crop is a mixed stand. Various forbs are blooming or ripening fruit from April until October. The same is true of the different species of grasses. There is a time for flax, another for mints, and still another for roses. Not all of these crops are bountiful every year. Some may form no viable seed. Indeed there is no pressing need among the perennials for fruiting each year. Only in wet years does little bluestem normally fruit abundantly on dry uplands. How unlike the more delicate, annual crops of man. Neither is there a critical time for drought as in wheat, or corn, or clover, where a few days of adverse conditions may prove disastrous. Height and density of cover vary annually as does also the tonnage of hay. If drought comes early, growth is resumed upon the advent of wet weather. If it comes later, the earlier yielding prairie components have had a good year. Nature's crops are adjusted to fit into periods of dry cycles as well as wetter ones. These have recurred again and again throughout the centuries. Reserves of food of native plants are extensive and their resources for obtaining water excellent.

Nowhere in North America is the soil more favorable for crop growth than in the tall-grass prairie, although sufficient water often becomes a limiting factor, especially in the western part. It seems clear that a knowledge of the degree of departure, in growing each kind of crop, from the mean established by Nature is highly desirable.

A NOTE ABOUT THE AUTHOR

Dr. J. E. Weaver is Professor Emeritus of Plant Ecology at the University of Nebraska.

Among the many professional offices he has held and awards he has received are these: President of the Nebraska Academy of Science, President of the Ecological Society of America, Honorary President of the 1950 International Botanical Congress; recipient of the Botanical Society of America's Certificate of Merit and of the first Nebraska Range Management Award, and designation for twenty-seven years by *American Men of Science* as one of the 100 outstanding botanists in the nation.

Dr. Weaver received his Bachelor of Science and Master of Science degrees from the University of Nebraska in 1909 and 1911 and his Ph.D. from the University of Minnesota in 1916. Before joining the faculty at Nebraska 48 years ago, he taught at Washington State College and at the University of Minnesota. He has also been a research associate for the Carnegie Institution of Washington, on the editorial board of the scientific journal *Ecology,* and a member of the National Research Council's committee on the ecology of grasslands of North America.

The author of *Vegetation of Nebraska* has written or co-authored over 100 articles. This is his fourteenth book.